P9-DIA-861

Praise for *Gifts of the Crow*

"Crows are amazing birds. They're smart, crafty, emotional, inquisitive, and wise, and form complex social relationships with other crows and a wide variety of other animals, including humans. *Gifts Of the Crow* is a wonderful, informative, and insightful book, well-documented and an easy read. It's the best book to date on these iconic beings."

—Marc Bekoff, University of Colorado Boulder, author of *The Emotional Lives of Animals, Wild Justice,* and *The Animal Manifesto*

"Throughout much of human history, crows have been our constant companions. In their exciting new book, Marzluff and Angell show us how crows' brains work, while providing the evidence that these cerebral birds have a lot more in common with us than we ever imagined. And Angell's illustrations alone make the book worth the price."

—Paul R. Ehrlich, coauthor of *The Birder's Handbook*

"*Gifts of the Crow* is a compelling book. Filled with wonderful stories of regular people's interactions with ravens, crows, and jays, it also cites engrossing scientific studies, reports on the field work of biologists, and offers detailed explanations of how the brain of a corvid actually works. I was fascinated."

—Suzie Gilbert, author of *Flyaway: How a Wild Bird Rehabber Sought Adventure and Found Her Wings*

LOVE

GIFTS OF THE CROW

How Perception, Emotion, and Thought Allow Smart Birds to Behave Like Humans

John Marzluff and Tony Angell

Illustrated by Tony Angell

Free Press
New York London Toronto Sydney New Delhi

fP
Free Press
A Division of Simon & Schuster, Inc.
1230 Avenue of the Americas
New York, NY 10020

First Free Press hardcover edition June 2012

Book design by Ellen R. Sasahara

Manufactured in the United States of America

1 3 5 7 9 10 8 6 4 2

Library of Congress Cataloging-in-Publication Data

Marzluff, John M.
Gifts of the crow : how perception, emotion, and thought allow smart birds to behave like humans / John M. Marzluff and Tony Angell.
p. cm.
Includes bibliographical references and index.
1. Corvidae—Psychology. 2. Corvidae—Behavior. I. Angell, Tony. II. Title.
QL696.P2367M357 2012
598.8'64—dc23 2011049130

ISBN 978-1-4391-9873-5
ISBN 978-1-4391-9875-9 (ebook)

To crows, so often maligned,
and
the people whom they engage, so often ignored

Contents

Preface

A BLUE-BLACK CROW PERCHES REGALLY on the cornice of a stone building on the University of Washington campus, where he is often found. Almost hourly, he delivers food to his mate and three fledglings, while also keeping watch for any threat to the nest. Suddenly he turns his head, caws softly, and glides away, landing on a lamppost directly above a blonde woman. The woman, Lijana Holmes, smiles and calls him "Bela" as she offers him a breakfast of eggs and meat, which she prepares daily. Bela, in turn, presents his special gift—recognizing Lijana and participating in this routine with her. His gift to Lijana is more abstract than what he provides his bird family, but it is powerful nonetheless—it is the ephemeral and profound connection to nature that many people crave.

Bela gives a slightly different gift this morning to my team as we walk through the same campus. For Bela knows us, and we know him. Five-and-a-half years ago we captured Bela and affixed light plastic rings to his legs for identification. So whenever he sees us, the old crow cocks his head, stares, takes flight and swoops low—right at us—screaming a harsh call that we immediately recognize as a bird scold. His family and neighbors hear the cry and join in, flying toward Bela to support his attack, and soon they, too, share his rage. The mobbing crows circle and scream above our heads just as they would do to a predator. Bela's discriminating actions give us remarkable and invaluable information, proving that crows can recognize and remember human faces. We wonder when, or if, he will ever forget (or forgive) us.

The gifts of the crow are physical, metaphorical, and far-reaching. Some, like Bela, provide understanding and companionship. Others have delivered sparkling glass, plastic toys, and candy hearts to their human benefactors. Some have dropped from the sky and shocked strangers by saying, "Hello." A raven, with its natural curiosity and conspicuous manner, can lead a hunter to game or alert a search party to the whereabouts of an injured person. A magpie or jay can brighten a cold day by pecking softly at a window to beg for its daily ration of food.

These birds are corvids, members of the avian family Corvidae, which includes nutcrackers, jays, ravens, magpies, and crows. We will consider many of the gifts with which corvids enrich the lives of people and the action of nature in the chapters ahead, and we will argue that a corvid's ability to quickly and accurately infer causation is itself a natural gift. It has survival value. This and other demonstrations of its mental prowess are gifts that all birds—and most likely their dinosaur ancestors—gained through evolution.

Crows' close association with humans has inspired art, language, legends, and myths. Corvids have their own form of eloquence as they exercise mischief, playfulness, and passion. They also lead us to reflect on their common behaviors with us and other sentient creatures and empower us with a deeper understanding of nature.

People from all walks of life eagerly recount the antics of their former pet crows or enthusiastically tell us authors about the fascinating, sometimes troubling behaviors perpetrated by their local jays, magpies, and ravens. In this book we celebrate their accounts along with others we have found in the scientific and popular literature, because these rare and exceptional behaviors cannot be limited to the few specialized researchers who study corvids.

Some scientists are dismissive of citizens' reports, viewing them as unreliable or unexplainable, because of laypeople's lack of formal training, lack of documentation, overinterpretation, and uncontrolled influences. To be sure, we have encountered descriptions of events laden with hyperbole and seasoned with more imagination than fact, but we were compelled to investigate them nonetheless and to interview the people who made the observations in order to verify the events. Taken individually, such stories are anecdotal, but collectively they provide a unique body of information that stimulates scientific exploration and becomes an assemblage of possibilities.

We draw from this cross-cultural collection to offer many intriguing stories about corvids' fascinating behaviors as we explore the anatomy and physiology of the bird brain. We have tested these anecdotes, such as those of the crow that summoned dogs or the ravens that windsurfed. Putting them through the scientific process, we evaluated each report for believability, precedence in the scientific and cultural literature, and the mental ability a bird would need to act in such a manner. We came to know the bird and the citizen scientist behind the observation as we examined as completely as possible what causes people and birds to share such poignant moments.

We recognize the intelligence and adaptability of this unique group of birds and base every thesis about their humanlike behaviors on how the brain of a bird is known to function. Through brain-scanning technology, which allows us to see within the crow's gray matter, we first glimpse how a crow's brain works through a problem. To date, most of the understanding of the inner working of the crow brain was derived from what was known from a few mammals and detailed investigations of song-learning in birds. We hope you will find, as we have, that understanding some of the neurobiological processes of crows adds mightily to your appreciation of how these remarkable creatures operate so successfully in our dynamic world.

Betty the New Caledonian Crow makes a tool to retrieve food.

1

Amazing Feats and Deep Connections

BETTY, A New Caledonian crow, peers briefly into a tall, clear vertical tube at the small basket of food inside it and pecks quickly at the plastic to test if she can break through it. Her brain is lit with electrical and chemical energy as she contemplates the puzzle of reaching the treats. Drawing on her experience in the wild using stems and branches, she grasps in her beak one end of a straight length of wire that researchers have left nearby and plunges the business end down into the tube in an unsuccessful attempt to spear the food. When that doesn't work, she tries to edge the food up and out by pressing the wire against the walls of the tube. After that doesn't work, Betty next fiddles with her wire and bends it, turning it from a spear into a nicely serviceable hook. With this manufactured tool in her beak, she deftly fishes the handle of the basket and lifts the prize from the cylinder just as a person might heft a bucket full of water from the bottom of a well.

Betty's fabrication of her fishing tool would easily best the efforts of other creatures that are widely considered to be her mental superiors. A dog, even a pooch of the cleverest breed, would probably paw the tube, shake it, chew at the top, and dig at the base and yet most likely fail to get any reward for its efforts. A human toddler also might rap on the container or try to muscle it over and eventually cry in frustration or pound the carpet with hands and feet. A concerned parent or older sibling might respond to the toddler's tantrum, reinforcing the child's inadvertent use of his social power and connections to acquire the prize and also perhaps reinforcing a learned helplessness or dependency. For this particular test, the crow was smarter than a dog or human toddler. But if we measured the ability of dogs, crows, and toddlers to learn voice commands, the dog would rule the day. Demonstrating mental prowess or the intelligence of a species requires a wide range of tests and knowledge of the animals' ecology and physical ability. Dogs may fetch newspapers and retrieve ducks for hunters, but without opposable thumbs they can't hold wire tools. Wild New Caledonian crows, on the other hand, regularly use their beaks to fashion hook tools from plant materials. This ingenuity—crafting a hook from a foreign material and using it to gain an unreachable reward in a new setting—requires thinking, appraisal, and planning, mental attributes that have rarely been associated with birds, until recently.

Most people consider birds to be instinctual automatons acting out behaviors long ago scripted in their genes, but *Gifts of the Crow* celebrates the fact that some birds—particularly those in the corvid family, which we generally call "crows"—are anything but mindless or robotic. These animals are exceptionally smart. Not only do they make tools, but they understand cause and effect. They use their wisdom to infer, discriminate, test, learn, remember, foresee, mourn, warn of impending doom, recognize people, seek revenge, lure or stampede other birds to their death, quaff coffee and beer, turn on lights to stay warm or expose danger, speak, steal, deceive, gift, windsurf, play with cats, and team up to satisfy their appetite for diverse foods whether soft cheese from a can or a meal of dead seal. You can think of these birds as having mental tool kits on a par with our closest relatives, the monkeys and apes. Like humans, they possess complex cognitive abilities. In fact, they have been called "feathered apes."

Some of the crow's achievements we discuss are well documented by credible observers. Most are plausible, albeit astonishing. And because these birds often live near people—in our gardens, parks, and cities—they often involve humans in their daring, calculated, emotional, and bizarre activities. Keenly aware of our habits and peculiarities, they quickly learn to recognize and approach those who care for or feed them and to avoid and even scold people who threaten or harm them. This important skill was learned and honed over millennia and enables crows to exploit friendly people and other animals and avoid the dangerous.

These birds are noteworthy in their ability to thrive in the midst of environmental challenges that discourage and even extinguish other species. The few exceptions are species that are isolated on islands: the Hawaiian crow and the Mariana crow, for instance, which are among the most endangered species on our planet. Other island crows, the majority of which are specialized members of diverse tropical-forest ecosystems, have also become rare; the Banggai crow of Indonesia, for instance, was even thought extinct until glimpsed in 2007 on small Peleng Island by intrepid local biologists.

The success of most crows has a lot to do with their flexible and complex social lifestyles, long life spans, and large brains, which are able to integrate and shape what they sense into reasoned action. In fact, corvids and our own species share qualities, which have enabled both of us to adapt to changing environments and to flourish.

The sophisticated behavior of crows overturns any lingering notions of a "birdbrain" as being unintelligent. Biologists have documented and explored corvids' language, delinquency, insight, frolic, passion, wrath, risk taking, and awareness. In *Gifts of the Crow,* we explore each of these topics to learn how crows' individual and collective social learning abilities enable them to craft tools, communicate subtle messages, plan for the future, intuit solutions, deceive others, and carefully adjust their boisterous lives to our unpredictable human nature.

Science is far from understanding the ways our own brains, let alone a bird's, produce complicated behaviors. But in the last two decades, our understanding of how a bird's cerebral equipment is organized, and how bits and pieces of it function, has made enormous strides. Now, we gain new insights every day.

The complexity of a bird's brain becomes evident when we examine its many parts, which we do throughout *Gifts of the Crow,* in Tony's original illustrations. Even phrenologists from the nineteenth century were enamored with the bumps and dents in the crow skull to which they attributed centers for caution, destructiveness, musicality, and imitation. The crow's brain is indeed packed with ability, but our modern scientific view looks deep within the brain. There, relays from the eyes, ears, mouth, and skin transport their view of the world to the brain stem and into the brain, which uses emotion to integrate and shape diverse information and past experience to guide the crow's behavior and enhance its survival and reproduction. Its brain allows the bird to learn quickly, to accurately associate rewards and dangers with environmental cues, and to then combine what it knows with what it senses and to draw conclusions leading to a more informed response.

Consider our drawing of the crow's brain-activity centers, which shows capabilities that we humans once thought were unique to us.

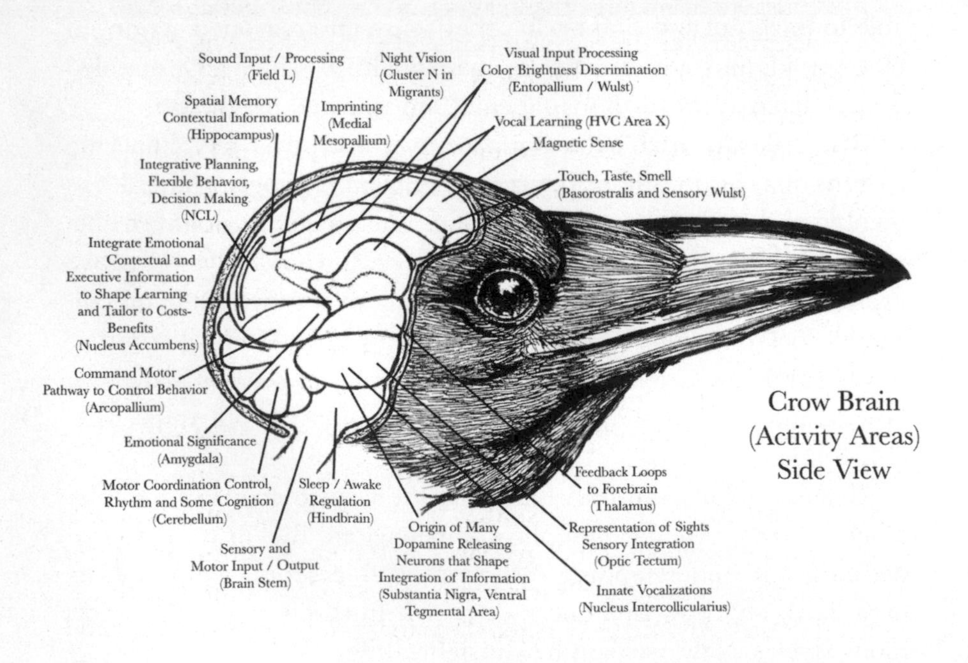

Crow Brain (Activity Areas) Side View

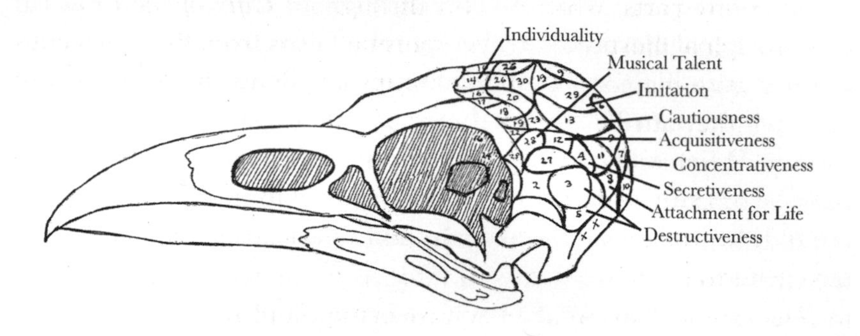

The Crow's Phrenological "Bust"

The complexity of the bird's brain is evident when one considers some of the many interconnected and integrated functions preformed by its various regions. Information from the outside world enters the brain through the brainstem, hindbrain, and midbrain. These regions are critical to maintaining basic biological functions, but also exert considerable influence over other parts of the brain when synthesizing and releasing a variety of chemicals—including hormones, opioids, and dopamine. The forebrain of a crow assesses sensory information, integrates this view of the environment with context and emotion to form memories, and sends electrical and chemical instructions to motor control regions to command action. The phrenological bust is an early view of the crow's brain from 1836. Rather than the distinct actions suggested by this image, modern views of the crow's brain (see previous page) emphasize how different regions interact to affect behavior.

In exploring the neurobiology of crows, we tell a lot of stories that show that humans and crows have an ongoing connection, a cultural coevolution, that has shaped both our species for millions of years. Archeological evidence reveals that our earliest human ancestors shared the company of ravens and crows. Early hunter-gatherers noticed these birds and celebrated them in legends and myths around the world over thousands of years. Ancient spiritual connections are evident on the cave walls of Lascaux, France, where a crow-headed man has been interpreted as the soul of a fallen hunter. The writings of early Scandinavians celebrated ravens as useful informants. The First People of the Pacific Northwest saw them as creators and motivational forces, while in the oral histories of Eskimos, the abilities of ravens to prevail in lean times led these people to tie a raven's foot around the neck of each newborn child to assure the survival of the next generation.

The Haida people of the Pacific Northwest placed their deceased shamans on elevated altars adorned with carved ravens, which symbolized the priests' connection with the creator; there their bodies would be eaten by ravens, which would free their spirits and allow them to travel. In Tibet in a similar tradition, dead loved ones' bodies were cut up and placed on tower platforms where revered ravens and vultures could eat them.

Throughout the world the mysterious, ingenious, and sometimes horrifying ways of crows and ravens have dramatically influenced our language, music, art, religion, and popular culture. A melancholy Edgar Allan Poe was moved to poetic fame by a talking raven. Corvids' mysterious dark form appears in Alfred Hitchcock's film *The Birds* and Vincent van Gogh's final painting (*Wheatfield with Crows*, 1890). And of course contemporary film continues to employ the crow as a metaphor for the unknown.

From our interface with corvid culture, the English language includes words like "crow's nest," "crowbar," "crow's feet," "rave," and "ravenous" among many other terms inspired by corvids. Names on the land—more than four hundred in Britain alone, from high and lonely places like Corby's Crag to those commemorating past transgression and execution like Ravenstone—reveal the species' influence on our sense of place. Even our most personal adornments—clans and family names—like Bertram (*wise raven*), Crawford (*crow foot*), Ingram (*the Norse god Ing's raven*), and Corbin (*little crow*)—commemorate corvids. The bold demeanor of these common birds is reflected in mascots chosen for sports teams (e.g., Baltimore Ravens and Adelaide Crows) and names of rock bands (e.g., Counting Crows and Black Crows). Repulsion at the sight of corvids scavenging human corpses produced public outcries to rid London of ravens after the fire of 1666, shaped the masks worn by doctors during the plague, and likely fostered taboos and colloquial sayings about eating crow. Continuing what is likely a relatively recent ritual, the British Crown ensures that there are always six ravens in residence at the Tower of London, where a Royal Keeper of the Ravens tends the six birds plus reserves, whose tenure is said to prevent the fall of the empire.

Place names often reflect our associations with corvids. Crows Landing is in California.

As people were influenced by corvids, so, too, did people influence the birds. Our ancestors probably first encountered these abundant, bold, and wily scavengers where they fed together—messy places like animal kills, butchering rocks, and fish-cleaning sites—where the species still meet today. Over time the birds learned to exploit the agricultural resources our ancestors produced and the food supplies that we dried or stored. Crows scavenge what we provide, expand their ranges to colonize our cities, and track our farming practices. They adjust daily travels and migratory traditions to both avoid and exploit us as necessary. The supplements we provide—food, water, shelter, and safety—have increased the abundance, survival, and birth rates of many crows.

We human beings and our activities afford corvids numerous opportunities to innovate. They harvest tenderized, accessible roadkill and even place hard seeds in the road so our cars will crush the shells and make the kernels or nutmeats available. They use our garbage as food and nesting material. They incorporate our voices into their language. They compete mightily with us and routinely irritate us, but our attempts to dissuade them have favored the warier, more secretive birds, as have diseases we introduce that challenge their immune systems and weed the weakest from the flock.

Some people won't even consider that the eight traits we examine—language, delinquency, insight, frolic, passion, wrath, risk taking, and awareness—can be characteristic of nonhuman species. But the science of bird intelligence and cognition has interesting, real stories that should compel a full examination of such an inflexible, antiquated view of humans as the center of the natural world.

By observing nature, we can bring a sharper focus to the picture of ourselves and how we fit into the system of life. Watching corvids in particular reveals some of the complexities involved in survival and the strategies other species employ to make their way in the world. We've found that when we reflect over what goes on in the lives of these birds, they give us ideas, insights, aesthetic inspiration, and, perhaps equally important, fresh considerations for our imagination.

We are hardly the first generation to benefit from watching the behavior of members of the corvid family. The author Charles Dickens, writing during the golden age of natural history in England, kept ravens as pets and was so inspired by their behavior that he included a raven as a principal character in one of his early novels, *Barnaby Rudge*. He named the raven character Grip, after a pet bird that he had taught to say: "*Keep up your spirit,*" "*Never say die,*" and "*Polly put the kettle on, we'll all have tea.*"

Dickens's description of Grip suggests that he understood animal behavior and employed his knowledge pointedly. The raven becomes

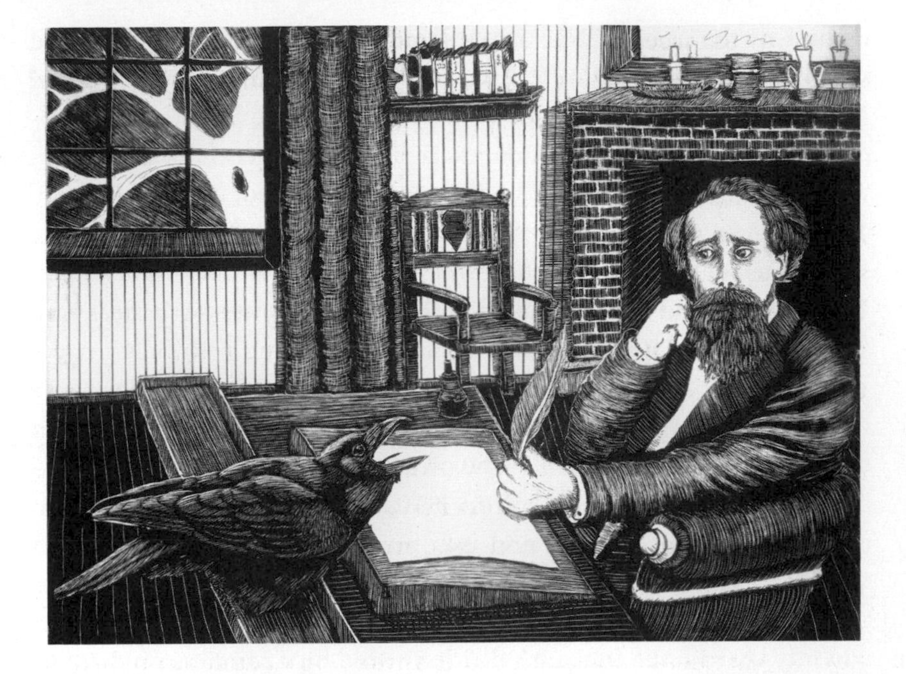

Charles Dickens consults his raven Grip.

the alter ego of the simple-minded Barnaby Rudge, who announces: "He's my brother, Grip is—always with me—always talking—always merry." Grip is referred to as "a knowing imp" and "sharpest and cunningest of all of the sharp and cunning ones." Time and again the raven advances the plot or reveals information when the human characters are unable to do so.

The inspiration of Grip the raven continued after *Barnaby Rudge* was published. Appearing in America in 1840, the novel was reviewed favorably in *Graham's Lady's and Gentleman's Magazine* by a critic from Philadelphia named Edgar Allan Poe. While he admired Dickens's novel, he wrote that the character of Grip the raven could have been employed even more successfully, had the author given the bird a greater prophetic presence. Four years later, Poe composed his famous poem *The Raven*. It clearly echoes a scene in chapter 5 of *Barnaby Rudge,* where, inside an English country inn on a dark night, an acquaintance of Barnaby's mother is attracted to a sound from outside and, thinking of the raven, asks: "What was that—him tapping at the door?" "No," replies widow Rudge, ". . . Tis someone knocking softly at the shutter." Poe expanded the use of the key word "tapping," changed "knocking" to "rapping," and intensified the prophetic possibilities of the bird.

While I nodded, nearly napping, suddenly there came a tapping,
As of some one gently rapping, rapping at my chamber door.
"'Tis some visitor," I muttered, "tapping at my chamber door—
Only this, and nothing more."

Grip died prematurely, and Dickens was so saddened by the loss of his companion that he had the bird stuffed. Grip remained in the family until the writer's death in 1870, after which items from his estate were auctioned off, and was purchased by Colonel Richard Gimble, a collector of Poe paraphernalia. Gimble later gave Grip to the Free Library of Philadelphia, where, to this day, nearly a hundred and fifty years since Dickens's death, the old bird remains on display, perhaps still bestowing some portion of its original gift of stirring the imaginations of those who see it.

2

Birdbrains Nevermore

A Clark's nutcracker with his pouch full of pine seeds.

A Clark's nutcracker named Hans, sharply dressed in bold gray, black, and white plumage, peeks through a portal into a small, featureless room in Russ Balda's lab on the campus of Northern Arizona University. Hans dives to the sand floor and digs a neat hole, his beak flipping sand until he finds a cache of pinyon-pine seeds. A month earlier, he alone had cached those seeds in this barren room. Now, hungry, he is deliberate and exceptionally accurate as he finds his seeds. He digs only where seeds are buried and recovers every one.

Over thousands of years, Hans's ancestors competed with the Anasazi, Hopi, and Navajo for this nutritious prize. A Southwestern tree of life, the pinyon adapted to having people and birds disperse its seeds. Birds and people learned when and where they could harvest seeds and how best to gather, transport, and store the piney propagules. The brains of Native Americans and nutcrackers expanded to remember what they learned. The rush for pinyon gold shaped body and mind.

Hans and Russ are pioneers in the study of avian cognition. Together, they showed the scientific world that an ordinary bird has an extraordinary brain. Nutcrackers and jays can remember for months the tens of thousands of locations where they cache pine seeds each year. This mental ability has enabled the birds to thrive in nature's boom-and-bust seed economy. The avian mind mastered the challenges of deep snow, long distance, time, and competition and was not often thwarted.

Russ loved to pit bird against graduate student to make his point. As one of his students, I, John, would have to bury nuts in the same room that Hans used. A month later my fellow students and I would return to the room to recall where we'd stashed each morsel. We found most, but not without a fair number of errors—digging where no seed existed or forgetting a cache entirely. Hans and Russ's other birds always kicked our brains. We grad students would assert that our accuracy would improve if we were given beer or money to cache, but when it came to pine seeds, the birds always won.

We human students prevailed in other arenas, however, and demonstrated how the über-social pinyon jay, a bird that lives in permanent flocks that may number hundreds of individuals, thinks about its social network. Pinyon jays do something that scientists previously ascribed only to monkeys—they infer their status relative to that of other flock members simply by watching others duke it out. The pinyon jays' complex society requires individuals to keep track of others' abilities, something that less social corvids are ill equipped to do. Knowing one's physical ranking saves a jay from a needless fight. Awareness of other facts about a flock mate might also be useful: remembering how each individual fared in raising offspring, for example, provides critical information to a recent widow or widower suddenly in need of a new partner. With this information, a previously successful parent can find a new mate of like ability. When a successful parental male or female jay loses its lifelong mate, it remates with another available and successful parent. Conversely, duds mate with duds. Living with others stretches the jays' minds, grad students' minds, and the minds of any social group.

On the other side of the continent, Bernd Heinrich studied ravens. In a simple aviary beside Bernd's Vermont house, food was hung on strings, and the grandest of all corvids retrieved the food by coordinating their gripping, pulling, and holding of the strings. Working with strings, the ravens seemed to be intuiting their way to a meal. After just a quick glance, they appeared to understand all the steps needed to retrieve the suspended prize. With Professor Heinrich, another of John's mentors, we students discovered some astounding abilities of ravens that allowed them to survive New England winters; when they find rich food bonanzas, they share them.

Ravens even understand the intentions of others of their own species. Ravens will move stashed food items when potential thieves who witnessed the original caching are present, and they will leave their stashes alone when around ravens who don't know that they've cached food. Ravens will also follow the gaze of a human to where foods are hidden just as a dog will. So ravens are also capable of some level of reading of other species' intentions and moods. Certainly they are students of human body language.

Heinrich's and Balda's successes at probing the minds of jays, nutcrackers, and ravens encouraged others to investigate the mentality of corvids. In England, in the laboratories of Nicky Clayton and Nathan Emery, western scrub-jays have demonstrated a refined sense of time and appear to imagine possible scenarios before acting. For example, scrub-jays recover perishable worms sooner than nonperishable seeds. Like Heinrich's ravens, scrub-jays also recache food items when pilferers can see or even hear them cache, which suggests that future rewards, not immediate hunger pangs, drive decisions.

Thinking about the future implies imagination, but mentally filling in the missing parts of a picture conclusively demonstrates it. Magpies are exceptionally imaginative. Like scrub-jays, these medium-size corvids remember what they did, and where, and they also can mentally represent a fully hidden object. German researchers played an extreme version of the shell game with black-billed magpies by hiding food under containers and moving the containers behind screens, and the birds were still able to track in their minds invisible objects among three different locations.

Scrub-jay deceives others while caching.

While animal behaviorists have chronicled the mental abilities of birds, neurobiologists have been peering deep into the mechanics of the avian nervous system—its labyrinth of nerves, spinal cord, and brain. In labs around the world, scientists scan whole brains using the same magnetic-resonance-imaging machines that help medical doctors diagnose disease in human brains. They section and stain brain slices and even wire up and record the electrical activities of individual nerve cells in order to understand how a bird senses and conveys the information flooding in from the environment and then how it assesses and integrates all this knowledge into a behavioral response. Even though neurobiology has made tremendous strides toward these ends, our understanding of how birds decide to act still remains rudimentary. We know more about how birds fly, migrate around the globe, and sing than we do about how they think.

A BIRD SENSES AND REACTS TO ITS WORLD

Imagine the world from a bird's perspective. Sounds that we cannot discern play in slow motion to a bird's musical ears, enabling it to discriminate messages hidden to us. Most objects loom large to birds' small bodies, but they can fly through, around, or over barriers, giving them unique perspective and the ability to explore fine detail. Their speed and agility make the living world seem slow, whether they are hovering to sip nectar, perching to spy a mouse, or sailing on a breeze as they eye a child fumbling with a sandwich. But, as a fellow vertebrate, a bird observes and reacts to its environment with much of the same equipment that we have. Birds don't refer to books, cell phones, or wristwatches as we do, of course, but their eyes, ears, mouth, nose, and skin transmit sound, light, taste, smell, pressure, and temperature to nerve cells just as ours do. These cells, or neurons, convert this information to electricity or chemistry with a physiological technology that evolved 600 million years ago (see Appendix, pages 211–213, for an illustration of the neuron and how it works).

The bird's sensory apparatus assesses its environment and conveys impressions that become a buzz of electrical pulses, which stream to the spinal cord for immediate reaction, or are transferred to the brain for central processing. As birds react, either reflexively and immediately or after some period of shaping by the integrative, memory, and emotional centers of the brain, a parallel set of nerves delivers electrical messages back to the muscles to control voice, movement, and general body functioning.

We will take a deeper look at what neurobiology has discovered about birds so that we can better put into perspective the fascinating behaviors of crows and ravens. And because we share an ancient common reptilian ancestor with birds, one who sensed the world much as we do today, we will also learn a bit about our own nervous system.

Neither we humans nor birds have to think about pulsing our hearts or digesting our food, because these jobs are done automatically in all vertebrates by nerve centers peripheral to our spinal cords and brains that join directly with our vital organs. Birds and people have some ability to control these actions because peripheral nerves connect to the spinal cord, and information from them is relayed to and from the brain stem and brain. But thinking before reacting takes time and energy, so basic life processes, many movements, and emergency responses are coordinated by our autonomic nervous system and its influence on our hormones and muscles. For example, when a robin detects a hawk streaking around your house intent on finding a meal, its autonomic nervous system stimulates the release of the hormone adrenaline, which revs up its heart and lungs to push oxygen-rich blood to feed skeletal muscles to act for a quick escape.

Watch a bird atop a tree swaying in a strong breeze. Simple nerve circuits between feet, muscles, and the spinal cord are automatically adjusting its body posture so that it remains firmly perched—all without conscious effort. Many other basic behavioral actions are controlled in a similar manner. Coordinating the various muscles that must be flexed and relaxed to walk and run on two legs, or take off, fly, and land, is done in the spinal cord. In fact, birds have special, enlarged junctions where the nerves controlling the wings and those controlling the legs join the spinal cord. These front and hind limb circuits are wired together by special nerve tracts in the spinal cord. The precise coordination of wings and legs allows birds to perform intricate behavioral displays and to move and work their lungs at the same time, something reptiles and amphibians cannot do. We provide an illustration of the bird's central nervous system in the Appendix, pages 214–215.

While wings and legs are mostly controlled by short neural circuits that do not involve the brain—which is why a chicken runs and flies even when its head is cut off—moving is not completely beyond thoughtful control. Special nerve tracts within the spinal cord carry neural messages from the wings and legs to the terminus of the spinal cord, the enlarged base of the brain known as the brain stem, and especially the cerebellum. In the cerebellum, that wrinkled ball that hangs like a ripe fruit just below the main part of our brains, independent processing of sensory information and its first integration with the complex forebrain takes place. To tilt a wing and bank left, the cerebellum modifies the electrical signals from the bird's spinal cord with input from the forebrain and then sends it back down the spinal cord to the muscles to adjust wing shape and position, or down the cranial nerves to coordinate head and eye movements. The cerebellum's shaping of electrical signals smoothes, coordinates, and calibrates the moves of people and birds. The important actions of the cerebellum can be disrupted by alcohol, for instance, which renders an intoxicated person unable to touch his nose, walk a straight line, or accurately steer a car. A bird can disrupt its cerebellum in the same way. We've seen crows gorge on fermented elderberries and wander into traffic or crash into buildings, drunk.

A bird's feet, feathers, bill, and skin are in touch with the environment and register many of the same sensations that human body parts do, via similar receptors. Sound waves bend hair cells in the ear. Light waves excite color and brightness receptors in the retina. Pressure waves impress nerve endings in the bird's skin; odor and flavor excite taste buds and cells in the mouth and nose. In addition, the chemical composition of cells in the bird's eyes responds to the angle of the earth's magnetic field.

The massive volume of raw sensory information streaming toward the brain cannot all be considered. Much of this is filtered away by a series of brain-stem relay junctions—intersections where many nerves converge. Filtering background chatter away from the brain allows animals to tune their attention to unusual and important sights, sounds, tastes, smells, and touches. Some of these sensations are novel enough or strong enough to discharge nearby nerve cells and send electrical energy rushing from the peripheral nervous system toward the bird's central nervous system—its spinal cord and brain—where perception is considered and tempered by emotion and memory into action.

The brain, of course, does much more than filter and nuance automatic actions. As a crow works among narrow spaces between the long rows of parked cars waiting to board the ferry and cross Puget Sound, for instance, we see a decisive brain in action. Here in the Pacific Northwest, crows have learned to take their scavenging walks in the intervals of time when the passengers are sitting tight within their autos awaiting the signal to drive to the boat. The bird's path is often festooned with the detritus of human meals and snacks. It approaches a wrapper, gives it a "lift" test to determine if there is food within, and promptly drops it if it's too light. By contrast, the bird strides by a metallic object, giving it nothing more than a sideward glance. Undeniably the bird is making decisions within its brain that are based on experience it has stored and refined as memory. But wait, there is more here. The crow gives each car to its left and right a quick glance, verifying its own safety and determining if there appears to be movement in its direction. A car door slams in the distance, triggering perhaps a memory of past dangers in these circumstances; the crow crouches, ready to fly. But nothing follows the sound to suggest any imminent danger to the crow, so it quickly resumes its insouciant walk up the open corridor.

The bird makes no attempt at stealth, walking where it can be clearly seen. It pauses from time to time to briefly study the people in the cars, surely watching body language to determine if there is something to fear. The crow appears to be measuring the possibilities of approach and whether it can gain something by it. A passenger who sees the crow rolls down a window, and the bird, having integrated many experiences of this behavior in this safe environment, stands its ground only a yard or so distant. Its decision, far from being random, proves an informed one as the passenger tosses a cracker to the panhandling bird. Success! The crow snaps up the food and flies out from the corridor of cars to the lower branches of a tree where it picks the gift into small, consumable portions. Should a cracker prove too hard, the crow will fly to the shore to soften—and perhaps flavor—it in the shallows. In another minute, with the cracker consumed, it's back at its work among the automobile passengers, making itself visible and looking for willing contributors to its welfare.

Crows looking for a handout in the ferry line.

EVOLVING VIEWS OF THE BIRD'S BRAIN

Before we dig too deeply into the inner workings of the bird's brain, we need some perspective on that brain's evolution and development. The complexities of brain biology have caused more than a few undergraduates to shift career paths, so we'll employ a common and apt metaphor as we present the science: the brain represented as an old-style filament light bulb.

The threaded base of the bulb represents the brain stem and cerebellum. The spinal cord attaches to the base, supplying some electricity to the convoluted basal filaments. Depending on inputs of power, filaments farther up in the glass bulb—the forebrain—glow and light the room. In the brain, the glowing filament doesn't just light the room; it sends electricity back down the cord to the muscles and sense organs from which it came. We were struck by this metaphor on a cold and wintry evening in Seattle as we turned a heat-sensitive video camera toward crows flocking to roost. The infrared sensor in the camera indicated which parts of the birds were using the most energy and, as a side product, producing the most heat. The massive flight muscles of their breasts glowed hot from working the wings. And the crows' heads were lit up like holiday lights. The energy of thought and coordination illuminated a darkening sky like strings of flying light bulbs.

When you look at the brain of a vertebrate—pick your favorite fish, salamander, snake, bird, mouse, or human—you will see striking similarities and subtle differences. In general, brain size increases in more recently evolved lineages, the birds and mammals. But the basic shape is immediately recognizable—a spinal cord that bulges to form the hindbrain and swells farther into a midbrain and thalamus; nerves joining these lower portions of the brain directly from the ears, eyes, and mouth; a prune-like cerebellum; and finally a forebrain. These structures, excluding the forebrain, all look and work pretty much the same in all vertebrates. The light bulbs are interchangeable because of a standard base. Evolution doesn't fix something that ain't broke, and our nerves, spinal cords, and lower brain regions have needed little fixing.

But if you look farther up the bird's brain at the forebrain, or cerebrum, you are confronted with the creative—and confusing—power of evolution, which modified the reptilian forebrain into functionally equivalent but different bulbs in birds and mammals.

Despite centuries of research, untangling just how the brain of a reptile evolved into the brain of a crow or human is challenging because the soft tissue of the brain does not fossilize. Because digging bones won't help their cause, neuroscientists and geneticists look for similarities and differences in the structures of adult brains and the instructions that produce them within the developing brains of fetal lizards, chicks, and mice. Gross changes in the location, appearance, and function of brain regions provide some rough answers, but because birds and mammals share a very distant—300 million-year-old—common ancestor and each line evolved from quite distinct reptiles, the fine details have been slow to reveal themselves.

All young vertebrates start life along nearly identical pathways. Within a few days, a fused sperm and egg cell divides and proliferates into a solid, then hollow, ball of cells. Early stem cells that could give rise to the outer, middle, or inner layers of the ball acquire a more scripted destiny. Some in the outer layer, or ectoderm, are destined to become the nervous system. In the week-old vertebrate, the nervous system looks a bit like a squashed bean. This rudimentary structure, called the neural plate, is essentially a two-dimensional map of the entire brain in intricate detail—one part will become the cerebellum, another part the midbrain, another the hypothalamus, and so on. The future complexity is all there, and the rules that transform the two-dimensional plate into the three-dimensional, interconnected brain are generally known. Specific proteins produced by genes in particular places of the developing embryo act in an ordered way to guide development. And this is the same for a mammal, bird, or reptile (essentially it is the same for all animals more advanced than sponges). This specificity of fate is key to understanding what parts of the reptile brain begat what parts of the brain of mammals, like us, and birds, like crows.

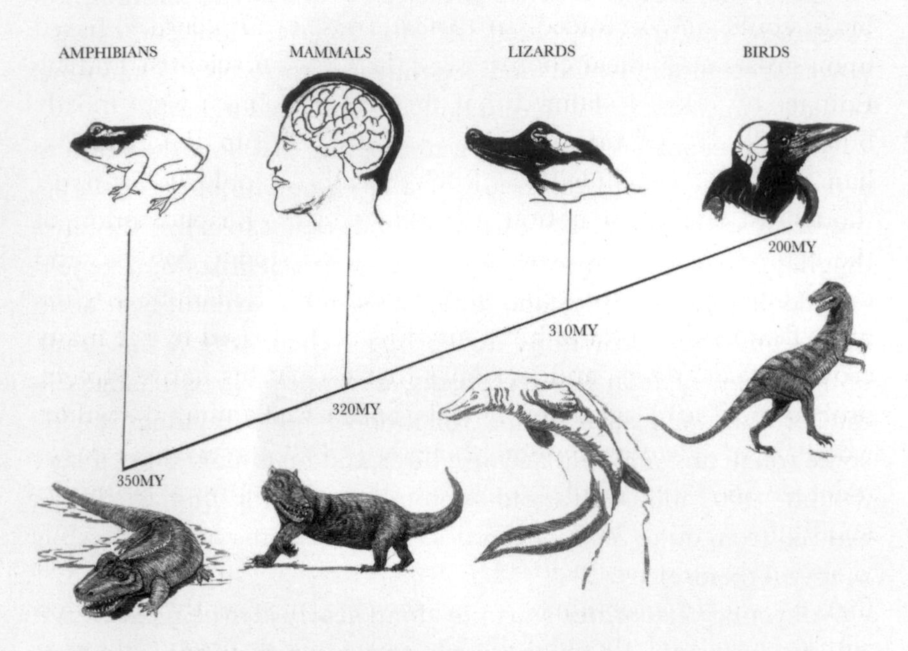

A dig about 350 million years back into the fossil record reveals the evolutionary history of vertebrates, starting with our common amphibian ancestor. Birds and mammals shared a common ancestor that lived about 300 million years ago, making the study of brain evolution challenging. Distinctive mammals first appeared about 200 million years ago and were derived from synapsid reptiles. Birds evolved only 160 million years ago from a different, highly divergent branch of the reptile family tree, the theropods that flourished as dinosaurs. Ancestral forms are diagramed below the evolutionary tree, or cladogram. Time is represented horizontally on this diagram, generally increasing from the common amphibian through the reptilian ancestors and from the slanted line representing this evolution up to modern forms of amphibians, mammals, reptiles, and birds. The diversity of size and shape of modern tetrapod brains is shown for a frog, human, crocodile, and raven.

Until recently, the genetic legacy of various parts of the bird's brain could not be traced. In the late nineteenth century, based upon gross anatomical comparison, the German scientist Ludwig Edinger mistakenly claimed that the brains of birds were mostly built from the primitive layers seen deep within the mammalian brain. He concluded that such structures could not support thoughtful actions, a notion that matched his misperception of the behaviors of birds, which he considered stodgy, robotic, and largely instinctual. He believed the behavior of mammals to be more flexible than that of birds and was unimpressed by the many crows, ravens, rooks, and jackdaws that walked his native streets, probed the nearby agricultural fields, and soared among the wilder woodlands. Yet the purposeful, thoughtful behaviors we have documented would have challenged his view of avian brains as primitive and limited organs.

Edinger and other neurobiologists who followed him observed that the smooth forebrain of birds—the part consistently associated with sophisticated, thoughtful behavior—had little obvious internal structure. The histological stains that were available to early neuroscientists colored the inside of a pigeon's brain uniformly. Even under the microscope, the stained pigeon brain seemed to be made up of the same type of cells that comprise the portion of the mammal forebrain that is below the convoluted gray matter, or neocortex. This lower part of the forebrain is crucial for motor control as well as the ability to improve movements through practice. It is composed of nerve centers collectively called the basal ganglia, which includes structures such as the striatum and globus pallidus. The neocortex, on the other hand, is derived from the pale, cloaking outer cover of the cerebrum known as the pallium. To Edinger and his intellectual descendants, bird brains appeared to lack a pallium and interconnections among brain regions that allowed thought, remembering facts and events, and reasoning. So, they concluded, birds could only express instinctual actions. To them, it was as if the bird's light bulb was full of filament, but the entire filament was made of the same material.

Konrad Lorenz, a pioneer in the study of animal behavior, was born fifteen years before Edinger died and grew up a mere four hundred miles from Edinger's Frankfurt, Germany, in Vienna, Austria. Lorenz's schooling reflected Edinger's contributions to the understanding of evolution and brain development. He would eventually share the Nobel Prize in medicine with Nikolaas Tinbergen and Karl von Frisch for their research that confirmed many of Edinger's beliefs. For instance, these great scientists showed that very complex behaviors of bees and birds could often be explained by instinct or imprinting, learning that occurred for a brief moment early in the animal's life. Ducklings learned whom their mother was simply by following the first thing that moved around their nest. Jackdaws, a common European corvid, instinctually mobbed owls or any animal caught possessing a dead bird. They even mobbed Lorenz when he carried a wet, shiny, black bathing suit that resembled a dead corvid. To Lorenz and the other pioneers, complex behavior did not require thought; it occurred whenever a simple stimulus in the environment was encountered. By providing the stimulus, or a mimic of it like the bathing suit, experimenters could trigger a bird's simple brain to release a complex behavior.

While jackdaws confirmed many of Lorenz's ideas, ravens provided an important challenge. Lorenz lived with a menagerie of wild animals, including a group of ravens. The dominant male raven was particularly fond of Lorenz and could not be tricked into mobbing a pair of black swimming trunks. Even holding another raven in distress would not guarantee mobbing. The male raven ferociously attacked a person who restrained his mate but flew up to Lorenz's shoulder if he pretended to choke a known subordinate raven. Rather than mob, the raven used Lorenz to his advantage and poked at the restrained upstart. From these ravens, Lorenz learned that instinctual behavior could be modified with these birds' personal experience. Lorenz's association with smart, social animals led him to realize that many birds learned, played, and even used insight to solve new problems, be it walking down stairs or around barriers. His books for the general public and for the scientific community from the late 1950s to the 1980s taught that most organisms survived because of their ability to learn, and that a few groups of birds, notably the geese and crows, routinely modified instinct by learning in ways not unlike we humans learn.

Science changes slowly, although it has periodic revolutions that sweep away outdated theories, so the characterization of birds as simpletons persisted until the 1960s. Neurobiologists then discovered that the striatal portion of a mammal's forebrain could be identified with new stains and tracers that highlighted the occurrence of specific chemical messengers like acetylcholinesterase and dopamine. These chemicals are abundant in the striatum, where they adjust electrical signals within and between nerve cells, but not in the pallium. Injecting these chemical markers into a bird's brain and looking at the results under the microscope showed that the avian forebrain was anything but uniform, as Edinger and others had thought. Rather, as in mammals, it had a deep, inner core of striatal tissue topped by a much larger heap of pallial tissue. The deep core is homologous—derived by descent from a common ancestor—to the basal ganglia of mammals. This similarity in brain structure and chemistry clearly indicates that birds' mental capacities are more complex than earlier scientists gave them credit for.

How the pallial tissue of birds and mammals evolved is more puzzling. There are big differences between this part of the brain in the two classes that were once thought to represent cumulative improvements in the "higher" evolutionary class of mammals. In mammals, for instance, the neocortex consists of six layers with distinctive cells and intricate interconnections that enable an integrated perception of the world. Adult birds, in contrast, have a pallium consisting of a small, three-layered "hyperpallium" and a larger area comprising distinct regions: the nidopallium, the mesopallium, the entopallium, and the arcopallium (we illustrate the basic parts of the bird's brain in the Appendix, pages 216–217). The bulk of a bird's forebrain functions like our neocortex, but its connections are among distributed, patchy centers rather than layers. Scientists agree that the neocortex of mammals and the hyperpallium of birds are homologous, both being derived from the upper surface of the ancestral reptile's embryonic brain. Some scientists, who are impressed with the similar neural wiring that connects and distributes sensory information from the lower reaches of the brain through the mammalian neocortex and the avian pallium, suggest an even greater homology. Embryology, though, is telling a different story.

To crack some ancient evolutionary riddles, one must look deeply into the developing embryo of a new life and retrace evolutionary history. The influence of our common ancestor is plain to see in the first weeks of embryonic development where fish, salamanders, snakes, birds, and people all look like tadpoles with well-developed tails, budding limbs, and rapidly differentiating brains.

In birds and reptiles, most of the pallium is derived from the lower surface of the roof of the fetal forebrain. In mammals the neocortex is definitely derived from the developing brain's upper surface. In a strict sense, then, the vast majority of the bird's forebrain is not homologous with the human forebrain. Our similar wiring patterns and mental abilities are the result of convergent evolution, where birds and mammals evolved large and interconnected forebrains from different basic parts of our common reptilian ancestor's brain.

This convergence is perhaps not so unlikely. Minor changes in the actions of a few powerful genes can construct radically different brains. In birds and mammals, the same genes in three critical regions around the embryonic forebrain produce proteins that proliferate and arrange neurons. It is as if the young forebrain is a battleground where proteins wage war for influence; in mammals, the victories occur on the upper surface of the pallium, but in birds victory is mostly on the lower surface. In each animal, a unique reptilian heritage comes to define what we recognize today as a distinctly mammalian or avian forebrain.

Our layered neocortex, once a signature of evolutionary "progress," is now known to result from a minor change in an ancestral gene's action. Layers arise in mammals when special neurons called Cajal-Retzius cells secrete a protein scaffolding along which newly formed neurons migrate over older ones to fashion successive layers, like the skins around an onion. The same cells work for a shorter period of time in birds to build a three-layer hyperpallium. Throughout the rest of a fetal bird's forebrain, Cajal-Retzius cells also lay out guiding latticework, but in the hyperpallium they create dispersed points, not radial zones, of attraction. Young nerve cells follow the proteins into clumps rather than layers. Why genes act on one bit of pallium more than another in birds and mammals may have a lot to do with the habits of our early ancestors. Early mammals were primarily nocturnal, so natural selection favored those able to sniff out, rather than see, a meal, mate, or predator. Expansion of the dorsal pallium in our ancestors may have facilitated their nocturnal life by enhancing their olfactory center and its ability to coordinate with other senses. The reptiles that continued to evolve after mammals diverged—those that gave rise to yesterday's dinosaurs and today's lizards and birds—were active during the day. In these animals, natural selection apparently favored those best able to discern foods, mates, and dangers with their eyes, not their noses. Ancestral birds whose brains derived more from the bottom surface of the embryonic pallium would match these criteria. Expansion of this ventral portion where visual, emotional, and auditory centers occur benefited a flying diurnal dinosaur and the birds that evolved from it.

THE AVIAN BRAIN ENGAGES WITH THE WORLD

How does a bird use its modified reptilian brain to sort through the confusing sights and sounds of a parking lot to find an easy meal? We can find some answers by tracing the actions of nerve cells from the brain-stem filters and relays up into the various regions of the forebrain and back again to the muscles that execute behaviors. Sensory information tends to flow in parallel streams to the forebrain, where it is influenced by other centers in the brain that are directly and indirectly connected. The form, strength, and route of the information's electrical signal may be changed before the streams are routed back to muscles and sensory organs.

Let's get into the interconnected structures of the corvid brain here, which will carry us back into more astonishing stories of these feathered apes.

The patchy pallium of the crow's forebrain works like our neocortex to assess and integrate the flood of information coming from its world. Some scientists suggest that either the extremely interconnected rear portion of the forebrain, the nidopallium caudolateral (or NCL for short), or an area toward the forebrain's upper edge (the dorsolateral corticoid area, or CDL), acts as the bird's executive center by selecting, scheduling, and adjusting neural responses from many sensory areas. Just as our layered prefrontal cortex orchestrates our behavior, so, too, do these areas of the bird's brain oversee its behavior. Individual neurons in the NCL of a pigeon, for example, integrate information about the amount of food and expected wait to get it. Their firing rates are high when the bird anticipates big food rewards, but they drop as the delay to obtain the feast drags on. A pigeon can decide when to take a small, readily available bit of food rather than continue to wait for a big feast by reacting to the activity of these neurons. But, first, sensory information from the outside world must reach deep into the most cognitive areas of the bird's brain.

Sensory information flows into the bird's brain through electrical coding of sound from the ear; color, pattern, and luminance from the eye; and smell, taste, and touch from the beak and face. All rush along cranial nerves through brain-stem relays to the cerebellum and especially to the thalamus. Some integration of sensory information occurs right away, in the midbrain, but most is sent forward to the thalamus, then relayed to specialized regions in the forebrain. Each of the important sensory areas, while sending information to nearby secondary processing areas in the forebrain, also has direct connections to the NCL. Electrical renditions of a bird's taste, sight, sound, and touch sensations all flow to the NCL. We provide a diagram of sensory flows within the bird's brain in the Appendix, pages 220–221.

This higher-level processing in the forebrain of a panhandling crow allows it to make sense out of the many sights, sounds, smells, temperatures, and pressures in a parking lot full of people and machines, allowing the crow to avoid danger and get food. This works roughly comparable to relays in a light bulb, with the analogy that distinct chains of filaments transport sound, smell, touch, taste, and sight through the base on separate courses through the more spacious glass bulb. The filaments converge on the back of the glass, and then the shaping of complex action begins.

BEHAVIOR SHAPED BY MEMORY AND EMOTIONS IN BIRDS

Bela, the discerning crow who works the grounds of our university, is guided by an upgraded version of a velociraptor's brain. He ignores most of the tens of thousands of people he confronts daily, their facial features filtered from serious consideration by junctions and relays in his lower brain. But he immediately recognizes a woman who brings nourishment to him predictably each morning. Rarely, no more often than once or twice a year, a researcher who long ago captured and tagged the bird shows his face, and again Bela immediately recognizes him as dangerous. Such perception, discrimination, and association are managed by the crow's brain. Bela's responses are not stereotyped or robotic; rather, they are nuanced to the possible reward or threat that each person represents, because his actions are informed by memory and charged with emotion.

Bela has formed and strengthened links between his nerves that perceive us and other places in his brain that contribute to the formation of memories and the feeling of fear. In humans, the amygdala acquires, stores, and links to other brain regions to express memories of fearful places and beings. In birds, this also appears to be the case. Bela's amygdala was active as we researchers held him, forming synapses with the forebrain's executive centers and the brain's memory stores in the hippocampus to cement a vivid memory of a dangerous situation.

The hippocampus is another brain structure that humans and birds share. But while the human hippocampus is buried deep within our forebrain, the bird's hippocampus is front and center, sitting atop the central part of each lobe of the cerebrum. Despite its different placement in birds and mammals, the hippocampus of both classes of animals plays critical roles in navigation, orientation, and spatial memory and provides important landmarks in mental maps that enable birds and people to travel. In both birds and people, the hippocampus of the left hemisphere seems especially important to forming maps and representing navigational goals. It is not simply the place that is mapped by the hippocampus; attributes of the place that may be important to contextualize future decisions are also mapped. Macaw, a raven that Tony raised, mapped his neighborhood as he explored widely from Tony's house. The raven's hippocampus no doubt facilitates Macaw's memory of favored places to forage, frolic, and avoid.

Memories are stored in the brain by its neurons. The neurons make memories either by changing their production of neurotransmitters or receptivity to them or when the neurons strengthen their physical connections. Russ Balda's nutcrackers and jays reinforced connections between neurons in their hippocampi in order to remember specific places. Special neurons, "place cells," occur in this part of the brain and fire when an animal is in a particular location of its environment. The neural connections are strengthened between the action of digging in the soil to stash a seed and the place cells that fire at that location, which enables the bird to mentally map its caches of food for future feeding. And when Bernd Heinrich's raven retrieved meat that was hanging just out of reach, it certainly increased the number and strength of synapses between neurons that fired to coordinate beak and foot motions so it could quickly gather in the prized treat.

Learning and memory occur because the brain forges strong connections among its various regions. As birds habituate to harmless stimuli, say a plastic owl atop a building intent on scaring them, their sensory neurons release fewer neurotransmitters so that reactive electrical spikes, which would sound the alarm at the sight of a real owl, are dampened and filtered from reaching the brain. In contrast, sensitization to dangerous stimuli, perhaps when a crow is narrowly missed by a gun-toting kid, causes normally inactive neurons to increase the production of neurotransmitters at the next sighting of a child with a gun. These short-term responses may become longer-lasting memories by changing the abundance of relevant synapses. Those synapses that were excited by the first sight of plastic owls may be pruned with repeated and inconsequential exposure, while those nerve cells excited by gun-toting kids may grow new connections with nearby nerves to more rapidly spread their excitation at the sight of potentially dangerous people. This integration of fearful stimuli occurs in the birds' amygdala or striatum, but these areas do not work alone. They are linked together with others and jointly integrate, assess, and respond to encountered dangers and rewards.

These connections are greatly influenced by chemistry. In places where important connections occur in the forebrain, nerve cells from the midbrain release the chemical dopamine, which enhances short-term memory that helps a crow survive. Memory of past decisions that produced favorable results, for example in a lab setting pecking a blue key and getting food, lasts longer when dopamine is present. Dopamine seems to shield engaged neurons from stimulation by other neurons, which would confuse memory and thought. Low levels of this important brain chemical in humans are associated with the inability of Parkinson's sufferers to coordinate thoughts and calibrate movements. The release of dopamine in the crow's brain likely underlies the crow's ability to quickly and accurately learn the cues to danger and reward provided by its environment that allow successful scavenging and hunting in places where few other birds tread. Other chemicals within the brain bulb influence the filamentous glow in the bird's light bulb.

Vertebrates and invertebrates have been shaping their sensory information with emotion and experience for millions of years. So why do some seem smarter—more nuanced and flexible in behavior—than others? At least two reasons stand out. First, animals vary greatly in the size of their brains, especially the size of the forebrain. Second, not all animals have the integrated looping of neural circuits between the thalamus and the forebrain that may enable actions to be reconsidered and related to new sensory input.

BIG BRAINS

In vertebrates, brain size tends to increase predictably with body size. This may be necessary for big animals to assemble, sort, and respond to the greater load of information coming in from their larger and often more distant eyes, ears, skin, and muscles. For any given body size, brain size follows another general rule: it increases from fish to amphibians, to reptiles, to birds, to mammals. Elephants and whales carry the largest brains, whereas the brain of a bass is hard to find. Of course, there are exceptions to the rule that brain size is in proportion to body size, because an opossum is small-brained while even small crows, parrots, and monkeys are remarkably large-brained.

We humans are the greatest exception yet; our brains are much larger than expected for a mammal our size. A human brain weighs about 1.3 kilograms (three pounds), or 1.9 percent of our total body weight. Earth's largest animal, the blue whale, has a brain that, while tipping the scale at 6 kilograms (thirteen pounds), accounts for only 0.01 percent of its body weight.

So, why do we get excited about a puny 14-gram (half an ounce) raven brain, or the 7.6-gram (quarter ounce) brain of a New Caledonian crow? Because, when standardized for their body size, the crow and raven brains are much larger than expected. In fact, as a percentage of average body mass, they approach or even exceed our brains; raven brains account for 1.4 percent of their body mass, while the New Caledonian crow's is a whopping 2.7 percent. Corvids in general have brains on par with similar-size mammals, not birds, and the crows and ravens in particular have brains the size of that of a small monkey.

The large brain size of a crow is not due to increased brain stem but is a direct result of its greatly enlarged forebrain, especially the mesopallium and the nidopallium. A crow's ability to adjust its behavior to new situations, such as the selection and modification of a new tool by the New Caledonia crow, comes from its prominent mesopallium. A European carrion crow's ability to consider an abundance of sensory information and devise appropriate actions is aided by its nidopallium, which is eight times as large as its brain stem. The nidopallium of a less cognitive bird like a quail or a pheasant isn't even twice as large as its brain stem, but parrots, especially large macaws, also have large nidopalliums. In fact, parrots' forebrains outsize even the biggest-brained corvids. The only other bird to challenge the corvids and parrots for the title of "feathered ape" is the black woodpecker. But as you might expect from a bird who pounds, chisels, and probes trees for a living, the mass of the woodpecker brain comes mostly from the coordinating and sensory center—the cerebellum, not the cognitive center.

The large corvid forebrain devotes more neurons and synapses to organizing and shaping a behavioral response to sensory information than that of any other animal, except parrots, monkeys, and some cetaceans (especially the smaller dolphins). These large brain regions have distinct subregions, because sending electrical and chemical messages across long reaches is less precise and takes longer than short-distance communication. This is likely why the nidopallium has distinct places for processing sound, touch, and sight, as well as executive centers that shape the bird's reactions. All of these substations are highly interconnected, which is why corvids have so many nuanced behaviors.

RECONSIDERING THE WORLD

When sensory signals arrive in the forebrain, memories and emotions shape them before passing them along so that they are different than when they arrived. This is called thinking. It tempers the raw sensory information and prepares new electrical signals to further influence thought and behavior. In birds, neural signals leaving the nidopallium go to the lower, rear portion of the forebrain, the arcopallium, which ushers electrical commands down independent, parallel circuits through the thalamus, midbrain, and hindbrain nuclei to muscle fibers whose actions create behavior. In the Appendix, pages 222–223, you'll see an illustration of how some adjusted sensory information destined to command muscles may actually return to the forebrain for further consideration by the thinking bird. A crow can use this neural loop between the thalamus and forebrain to reconsider its actions. The thalamus enables reprocessing of action commands from the arcopallium by restimulating the neurons in the forebrain that originally influenced the arcopallium. This mental loop, or reconsideration, strengthens important linkages between neurons to reinforce useful behaviors and provide a basis for improvement of other behaviors. These advances allow birds to sequence and learn complex activities.

The ability to reconsider an action by restimulating the neurons that produced the action is fundamental to conscious thought. Here is how it might work. A bird does not simply move through its environment sensing and reacting. The decision to move, for a bird, as for a mammal, is an integrated response to sensory, emotional, and remembered information. As a bird flies, it looks and listens. The sights and sounds stimulate its nervous system to produce electrical codes that instruct its body to reposition. Some of these codes pass through the thalamus and restimulate the forebrain. In the bird's brain the sights and sounds that caused it to move in the first place are experienced again. Now two sensations are stimulating the forebrain: one is the reaction of neurons to sights and sounds sensed by the eyes and ears; the other is the reaction of neurons to earlier sights and sounds that evoked behavior, perhaps adjusting a flight course. The two sensations are indicative of the present situation and an expectation of the situation, given the bird's past behavior. The difference between these sensations—a reality check, of sorts—can be evaluated by other brain regions sensitive to current emotional states like the amygdala, and perhaps by those with spatial information like the hippocampus. This continuous consideration of what they're sensing allows birds to think before and as they act.

How birds consider all they encounter is more complex than we have described; as with us humans, it is likely that all parts of the forebrain participate to perceive, integrate, and respond. The result is that thoughtful, not robotic, actions ensue. And because adjustments are continuously and consciously made, a bird like a crow, which has a large forebrain (for evaluating input) connected to a thalamus (enabling mental comparisons) can integrate what it senses, what it knows, where it is, how it feels, and what it expects into behavior.

Think again about our crow walking between cars at the ferry line. When startled by a car's engine backfire, the crow flies momentarily toward the sharp noise. This may seem counterintuitive, but he does this because a strong wind is blowing and, by flying into it and toward the same direction the sound came from, the crow is given immediate lift and more likely a quick escape from danger. The bird's brain combines new sensory information with experiences stored over a lifetime in a fraction of a second and creates reasoned action.

No wonder crows are so engaging. They think like we do. And we respond to them because we share kindred circumstances.

KEEPING THE BRAIN TUNED-UP

A crow of any age can add to its repertoire of tricks. Even the oldest crows that live among the tens of thousands of students at the University of Washington's Seattle campus quickly learn about new dangers and opportunities. These black-feathered practitioners of lifelong learning patrol the stately grounds, sizing up the next sucker who will share his lunch. In part this may be possible for a crow because, throughout its life, a bird's brain is dynamic, making new neurons, even as adults. This ability is rudimentary in humans.

Like an old dog that can learn a new trick, an old crow can form new connections among the neurons that respond when a different task is learned, and it can increase the ease with which the involved neurons fire. But it can also substitute new parts into the circuit. These fresh cells can be extremely effective in building even stronger, longer-lasting connections, essentially underwriting some long-term memory. New neurons and the genes carried within their nuclei may produce new proteins that strengthen or create different synaptic junctions that amend behavior. By replacing old neurons that have reinforced less important connections, birds can selectively change their brains to remember and coordinate an evolving battery of tasks.

The ability to rewire its brain may be important to the crow's ability to adapt quickly to changes in food supply. Old wiring that was tuned to raiding a farmer's field can be converted to enable feeding from a local dumpster if the farm is plowed under for a shopping mall and the dumpster becomes a routine fixture. Change in food availability modifies strategies and behaviors so that the crow can get at the different food supply. The crow's habitual caution around farmers in the field is no longer as relevant, so it learns to tolerate vehicle traffic. Would a country boy adapt as easily if thrust into the machinations of city life?

Brain space is limited, even in the voluminous forebrain of a crow. So, while the bird remembers more recent information, it loses some of the old information. In fact, this occurs seasonally. When nutcrackers and jays are busy each autumn caching tens of thousands of seeds that will sustain them throughout the coming winter and spring, the hippocampal region of their brains grow. Scientists are not sure what this means, but the new neurons that migrate to the hippocampus may become new place cells or fresh links to more recent memory. They may produce new sorts of memory-enhancing proteins or simply be more active than the old neurons they replaced. In any case they enable the birds to form a fresh mental map of caching locations. Nutcrackers may have little room to learn much else during this time, but they know where their cached seeds are, and that means survival. Russian birds have been reported digging accurately at the precise angle needed to reach seeds cached months earlier and then buried by a meter of snow. In spring, when other foods are gathered, the hippocampus and its outdated memories shrink. Places the birds no longer visit for food are essentially wiped clean from the mental map, and new neurons are readied to make different connections that will enable these remarkable birds to exploit the insects and small animals that are part of a summer's bounty. As the next crop of pine seeds ripens, the nutcracker's hippocampus swells to remember the locations of thousands of buried seeds—a real biological treasure map.

All this thinking doesn't come cheap. Although the brain of an animal is less than 3 percent of its body weight, the brain uses a disproportionate amount of energy. In humans, 20 percent of our daily energy intake goes straight to the brain. Given the relative size of crow brains, their maintenance is likely just as costly. Most of this energy maintains the slight electrical charge in the resting membrane of a neuron. Mental calisthenics, indeed! This high cost of a large brain cannot be borne by all animals and likely requires access to a high-quality diet, especially one rich in protein and fat, which provide the energy neurons need to move chemical and electrical signals around the nervous system. Humans' access to meat and seafood in the Pleistocene Era may have enabled our brains to evolve to their large size. Similarly, dining in tide pools, scavenging from the kills of predators, and specializing on oily pine seeds may have opened the door to an expanded corvid brain. In fact, those corvids that are scavengers or high-quality seed specialists—the magpies, nutcrackers, pinyon jays, crows, and ravens—outbrain their more vegetarian kin.

A rich diet enables a large brain, but evolution requires a benefit from increasing brain size. The ability to manufacture tools, use social information, adjust behavior to changing conditions, and track and exploit high-quality foods were probably important benefits that favored large brains in humans, as well as in corvids. With such returns, one might expect brain size to increase beyond a measly 3 percent of body size, but other structural constraints limit the upper size. Human mothers know what constrains our brains: birth. The need to stay light for flight likely also constrains a bird's brain size. Eating rich foods allows for a smaller gut for digestion, and the weight savings of a lighter gut may allow a large brain. Maybe the fast-food scavenging crows around the world's drive-ins are in the process of evolving even bigger brains considering the greater amounts of energy they receive from the fat-laden, processed items they regularly eat.

Maintaining a large brain requires energy as well as some downtime from constant stimulation, a break from the external world. Sleep provides this as sensory inputs and muscular responses are mostly silenced, even though the sleeping brain is anything but silent. Recordings of brain activity (those familiar brain-wave charts called EEGs) reveal a series of changes as we sleep. Our brain waves slow in frequency as we enter deep sleep, but they cycle predictably throughout the night from slow to rapid wave forms. We're mostly familiar with the rapid form, known as REM, or rapid eye movement, sleep, but both slow-wave and REM sleep are related to the maintenance of our cognitive brains.

Exactly how sleep maintains learning, memory, and reasoning abilities is the subject of debate and continuing research, but the ability to cycle between slow and rapid states of brain activity is important to a complex brain. The fact that REM sleep and coordinated, brain-wide synchrony of slowly oscillating brain activity are found only in birds and mammals—those animals that have the neural feedback loops between the thalamus and the forebrain that enable conscious consideration and reflection of their worlds—lends strength to the idea that sleep helps maintain complex brains in birds just as it does in humans. In birds this maintenance by sleeping is essential no matter how they get it, and in fact a bird is able to put one hemisphere of its brain to sleep while the other side works. This is especially important to migrating and oceanic birds who must sleep on the wing. That is an evolutionary advancement that our teenage daughters could also use.

In slow-wave sleep, where the signaling of neural circuits occurs less frequently, synapses in the forebrain uniformly lose strength. Reducing the baseline activity of all synapses saves energy and purges relatively weak synapses; their lowered activity doesn't release neurotransmitters that reinforce and build important connections. In contrast, the strongest synapses, those used most frequently or those between extremely well-connected neurons, are now more pronounced and don't have to compete for attention from the integrative centers of the brain. This relative strengthening of important synapses and pruning of less important ones creates new space on the neural slate where the coming day's lessons can be written. It may also explain why learned tasks are performed best after a sound sleep. It really does help to "sleep on it."

REM sleep may additionally shape information and memories that birds and mammals have learned throughout the day. During REM sleep the hippocampus and amygdala are active, and the output to muscles, except those of the eye, are completely silent. While appearing passive, sleepers mentally relive their experiences, igniting neurons in the forebrain to loop with the thalamus. The amygdala may add emotional significance to specific places as the hippocampal place cells fire. Connections are made between disparate experiences and charged with emotion. By mentally retracing and recompiling events, the significance of some are strengthened and that of others reduced so that, when REM sleep cycles back to slow-wave sleep, important synapses are made more so, and less important ones are lost. Memory is consolidated, adjusted, and fine-tuned for future purposes.

RECONSIDERING THE US AND THEM ATTITUDE

For centuries, many philosophers and scientists have believed that other animals were incapable of conscious thought and emotion. But as we learn more about how information travels through the brains of other mammals and birds, and how similar this trip is to the way our own brains work, we can no longer perpetuate this self-serving idea. Consciousness appears to depend on an integrative forebrain, and especially on its reciprocal connection to the thalamus. The connected loops of neurons that originate in the brain stem, pass through the thalamus, and course up to the forebrain before checking in again with the thalamus or commanding muscles are an important neural basis of consciousness. Animals with loops between the thalamus and the forebrain have expectations—in other words, they are able to consciously think. Birds and mammals have these loops. Reptiles' loops are minimal. Loops are unknown in amphibians.

A REFINED LOOK BACK AT BETTY THE CROW

We've covered a lot of ground, so let's reconsider Betty, the wire-bending, tool-wielding, New Caledonian crow again, now that we have an improved neuroscientific perspective. While a full mechanistic accounting of Betty's thoughts is beyond the capability of current scientific knowledge, we can better understand the sort of complex interaction between environment, neuron, and brain that must be occurring. Betty's retinal cells sense the room. They release glutamate, which paints in her midbrain an electrical picture of the food, the tube, and the wire. The electrical code for the image is relayed to her forebrain where she integrates this information with her memory of crafting hooks from the leaves and twigs of her native island and sees that wire is not dissimilar from the stems of leaves. She picks it up and her beak senses its flexibility. Memory and emotion command her muscles to bend the wire. She is planning, forming mental visions of success, and willing to wait for gratification. She first employs this wire "stem" as she would in the wild to stab and probe at her food. No luck! Now, perhaps she remembers bending a stem that felt like the wire or one that bent when it was stuck under a rock, so she wedges the wire into the duct tape at the base of the tube and walks the other end around, leverlike, to bend the straight material into a hook. Continual feedback from Betty's trigeminal nerve sensing the wire in her beak and from her optic nerve visualizing the making of the hook adjusts the commands to her muscles, and Betty expects a result. She thinks about whether the hook making is going as it should and adjusts as she needs to. She is relying heavily on the right eye–left brain visual system to sense her progress and on her cerebellum to guide her fine motor skills. She is solving a new problem by integrating sensory and memory electrical signals in her forebrain. Fishing the food from the tube is easy; with leaf-stem tools she has hooked grubs from under bark or deep crevices her entire life. But still the dexterity of her hook use is superbly guided by continual consideration of the expected and perceived neural signals in her forebrain.

We are not sure exactly when this experiment was done, but we'll put our money on it being shortly after a sound sleep.

3

Language

Crow lecturing dogs on the campus of the University of Montana.

Here Boy! Here Boy! The phrase was emphatic and clear. Whistles along with the words conveyed the caller's urgency and quickly attracted the dog's attention. It was early morning, and Vampire, a young dark German shepherd, was barking and lunging in her kennel outside the house, making a ruckus sufficient to rouse her owner, Kevin Smith, from a deep slumber. Kevin went outside and commanded Vampire to be quiet, but she ignored him. Strangely, so did the instigator who continued calling out to the dog. Kevin was preparing to reprimand his dog and confront an apparently presumptuous intruder into a cool Missoula, Montana, morning, when from behind Vampire's kennel bounded a crow that continued to call and whistle for the dog. Doubting reality, Kevin spoke to the crow that approached him, and the bird cycled through a series of well-rehearsed phrases. This crow had plenty to say, but raised more questions than it answered.

The crow left Kevin's property, but it didn't go far. It would become a well-known visitor to the nearby University of Montana campus for the next several weeks. There, the 1964 spring quarter was in session, and the talking crow was holding class on the university's central green, the Oval. Perched low on a branch of an oak tree, the crow called to its pupils—dogs of every breed, size, shape, and color. A pack of mutts focused their attention on the crow from the base of the bird's lectern. The crow had likely rallied them, as it had tried to gather Vampire, from the nearby neighborhoods and lured them to this learned spot. But why? The answer was suggested when the school bell chimed and the students spilled into the Oval, heading to their next classes. The crow took off low, only a few feet off the ground, with its devoted crowd of canines in noisy pursuit. In and out, the black corvine Pied Piper threaded a mayhem of canines through the students, creating confusion, wonder, and collision. When the students got to their classes, the dog-and-crow show stopped, and the bird again resumed lecturing from a low branch to its rapt class of dogs.

We have few further details from Montana because the talking crow disappeared as suddenly as it had arrived. We don't know whether the crow ever got a reward for its antics, perhaps a dropped sandwich or bag of chips from a startled student, or where it learned to talk. Did it know what it was doing? Why did the dogs stay? Perhaps the crow stopped appearing because it migrated farther north or got within reach of one of its pursuers.

Dogs following crow amid students.

It is not uncommon to have only an incomplete glimpse at the workings of nature, especially her most intriguing and rare actions. Anecdotes like Kevin's stimulate creative thinking and can guide further research. But blind acceptance of interesting stories can confuse our understanding of nature. It is hard to determine if such episodes are special or typical cases without knowing something of the observer, the conditions surrounding the observation, and how varying those conditions might affect the plot. Science teaches us to be skeptical, especially of the fantastic. So we immediately ask: Is Kevin trustworthy? Are talking crows typical? And, if not, why did this crow talk? Employing the scientific method allows us to address our skepticism. We pose several alternative explanations and confront these "hypotheses" with other observations. Our goal is to reject some explanations and thereby whittle many possibilities down to a few that are most likely. Three hypotheses are immediately relevant to the Missoula crow:

1. Kevin was dreaming, scheming, or prone to hearing voices.
2. The crow was an escaped pet whose owner routinely called his dog, and the crow had learned to mindlessly imitate the command.
3. The crow was an astute observer of human behavior who could not only imitate, but understand the concept of dog calling and employed the human language for personal gain.

We can quickly reject the first hypothesis. Our conversations with Kevin revealed him to be a reliable and well-trained observer. He holds a degree in wildlife biology, knows crows, and swears he wasn't drinking the night before the crow called his dog. Moreover, while amazing, a talking crow is not unusual. Corvids are famously loquacious, having extensive vocabularies of calls that convey explicit messages about danger, food, territory, identity, location, and emotion. They routinely mimic natural and man-made sounds like running water, honking horns, and crowing cockerels. And they often co-opt human words for their own use.

Halfway around the world and decades earlier in postwar France, another crow acted much like the bird from Missoula. A man's pet carrion crow, a close relative of the American crow, distinctly mimicked the voice of the lady next door. *Revenez ici* ("come back here"), the crow called to the young boy in the garden. Sure it was his mother calling, the obedient son ran to the house, but his mother was nowhere in sight. Another closely related species of crow from Austria, the hooded crow, lived to tell a tale. This bird, under the care of Nobel laureate Konrad Lorenz, returned home after a long absence, limping on an injured foot. Lorenz learned firsthand about the injury from the crow, which startled the scientist with a short German sentence translated as *"Got 'im in t' blommin' trap!"*

Other corvids also speak. Magpies, like crows, can say *"hello"* and often repeat their names. We've held conversations with Ted Turner's former pet, Harry, who now resides at the Red Lodge, Montana, Beartooth Nature Center. Regardless of what we say, Harry says only *no, no, no*. According to the folks at the Nature Center, Mr. Turner surrendered his magpie because it was aggressive, something that often makes keeping a wild pet difficult, although we also wonder if Harry may have been unduly critical of the communications mogul's ideas. Esther Woolfson, an English woman dedicated to the rehabilitation of many birds, often conversed with her pet magpie, Spike. Most of her conversations were decidedly one-sided, but not all of them. When Spike was missing, Ms. Woolfson would call his name. In response, from a hidden place, Spike would shout, *"What?"*

When ravens talk, even emperors listen. After defeating the joint forces of Mark Antony and Cleopatra VII at Actium in 31 BC, Octavian returned triumphantly to Rome to become the new empire's first emperor and acquire a new name, "Augustus." Macrobius Ambrosius Theodosius, an Italian philosopher and historian who held an important post in the imperial administration, recounted in the fifth-century book *Saturnalia* that Augustus met a talking raven. Parading through the streets of Rome, Augustus was congratulated on his victory at Actium by a man who held a raven. *"Hail Caesar, the victorious commander,"* croaked the bird. Delighted, Augustus bought the speaking raven for 20,000 sesterces, a considerable sum. But then a friend of the bird's suddenly rich trainer spilled a secret to Caesar that there were actually two birds. The second raven was brought forth and said as it had been trained, *"Hail the victorious commander, Antony."* The trainer had taken no chance in the outcome of the great war. Amused, Caesar commanded that the payment be split between the two men. This started a run on talking birds, and Caesar soon had parrots and magpies to keep his raven company and greet the emperor faithfully. Another raven raised by an enterprising, but poor, shoemaker seemed stubbornly mute, which caused the shoemaker to often lament, *"My effort and my money, down the drain."* Eventually the raven learned to hail the emperor, but when the shoemaker offered it to Caesar, he declined, saying he had an abundance of these greeters at home. At hearing the bad news, the raven spoke the other phrase he had heard often repeated, *"My effort and my money, down the drain."* Caesar laughed and purchased the astute raven, paying the highest price yet.

Ravens, as well as magpies and crows, learn their names and come when called, but they also seem to understand that their monikers symbolize their owners. A raven named Joe from Denver, Colorado, used his owner's name, Bob, as a greeting. *"Hello, Bob,"* the raven would say, until the day Bob could no longer care for the bird and donated him to the Denver Zoo. The former chatterbox fell silent, for years. But the raven did not forget Bob or speech. Years later, when Bob surprised his former pet with a visit, the raven immediately yelled, *"Hello, Bob."*

Tony's pet raven, Macaw, would greet him daily by calling, *"Hello, Macaw."* An important function of vocal imitation in birds is the maintenance of social bonds, so it is perhaps not surprising that Macaw would imitate a unique utterance from his important social companion. Konrad Lorenz named his raven Roah, a Germanic rendition of the typical call made by ravens. The bird discriminated between the name and the call, intoning Lorenz's *"Roah"* when communicating with the scientist, but croaking *roah* as would a raven, when conversing with wild birds. Roah used his label to steer Lorenz away from danger or when he desired human companionship, much as Macaw called Tony for a morning chat and the breakfast that usually accompanied it.

The ravens that reside at the Tower of London are perhaps the best-known and most closely watched corvids on our planet. These birds are commissioned privates in the British Army and are cared for by a corps of Beefeaters, under the command of a Royal Ravenmaster. They walk the Tower grounds and interact freely with legions of tourists. Their comfortable reign at the Tower is due to a myth that their presence there will ensure the British Empire's future. A more tangible result is that the ravens learn from people. And they are quick studies. One raven was seen perched on the shoulder of a tourist who wouldn't share his lunch with the bird. The raven whispered, *"Keep to the path,"* and the startled man wheeled around only to see the raven drop swiftly onto his lunchbox and make off with the food. Another of the birds, a large male named Thor, would sometimes correct Ravenmaster Derrick Coyle when he would say, *"That's for you"* as he fed Thor. Thor sometimes replied, *"That's for me."* It is impressive that Thor appears to grasp the nuance of human grammar, modifying a familiar, often-repeated phrase to better suit his needs.

All the cases of talking crows, magpies, and ravens that we know of involve pet birds. This is consistent with the basic premise of our second hypothesis that the dog-calling crow from Missoula was an escaped pet. We have shouted our share of insults and questions to thousands of wild corvids, and to date, not a single one has responded in "words" we can understand, although they mob a short distance from us in a behavior that suggests their ire and distain. Pet crows, on the other hand, can easily be trained to talk, since they are surrounded by human language and focused on human caretakers. If people raise wild nestlings, the young crows may even identify themselves as humans, not crows. Some birds learn what they are by the process of imprinting, where the developing nestlings establish their identity by seeing the adult who provides it food, warmth, and shelter. Words, phrases, or noises that are often repeated around a pet crow are quickly included in their vocabulary, accurately learned, and rarely forgotten. But how do they physically manage to speak human words?

Try calling a dog without moving your tongue or lips. Unless you are a ventriloquist, the dog is likely to look at you with a confused, pitiful expression. Yet talking crows mimic precisely the nuances of our speech simply by adjusting the muscles within their throats. Crows don't have lips as we know them, and their tongues are fibrous, not fleshy and muscular like ours. The tip of a crow's tongue is like a plastic spear, useful to sort the edible from the inedible, but not to form words. Our lips and tongues are critical to enunciation and give us final control over what we say and how we say it, but not so for a crow.

Pliny the Elder, a Roman from the first century, puzzled over the abilities of talking birds. He knew of magpies, crows, ravens, and parrots that spoke, often in full sentences. He was told that magpies so loved to speak that when they could not remember a cherished word, they died. He noted that most speaking birds had broad tongues and that the especially fleshy tongue of a parrot might account for its nuanced utterances. The talking raven, with its more fibrous tongue, appeared to speak from the throat, not the mouth.

Pliny's insight was accurate, even though it was based on incomplete information. The speaking ability of birds comes from four lips, called labia, buried deep within their throats. These fleshy blobs of connective tissue are puckered by the actions of six pairs of muscles, and they are housed within a special voice box called the tympanum. The whole apparatus, or syrinx, was a key breakthrough in the evolution of birds that endowed some species with the gift of complex song. Crows, ravens, and magpies are part of the lucky lineage of songbirds. They share a common singing ancestor with more notable singers like canaries, mockingbirds, warblers, and thrushes.

What is really neat about the syrinx is that it allows birds to sing from two mouthpieces at the same time. One pair of lips within each of the two bronchial tubes that connect the crow's lungs and the air they expel to the base of the syrinx vibrates to produce crow sounds. When compressed by the syringeal muscles, air that once flowed uninterrupted, like water around an island, is now constricted just upstream of its confluence and forced through regulated right and left channels. The vibration of the lips produces the sound we hear. The degree to which the syrinx moves forward and compresses the flexible trachea controls the pitch and intensity of sound. The mixing of sound from the right and left "mouth" produces the incredible harmonic structure characteristic of birdsong. When you slow down a recording of a thrush's song, you can pick out two distinct melodies that the thrush sings simultaneously, as if singing fine harmony with itself. Our fast tongue in a flexible mouth is really no match for four lips, two mouthpieces, and six pairs of syringeal muscles.

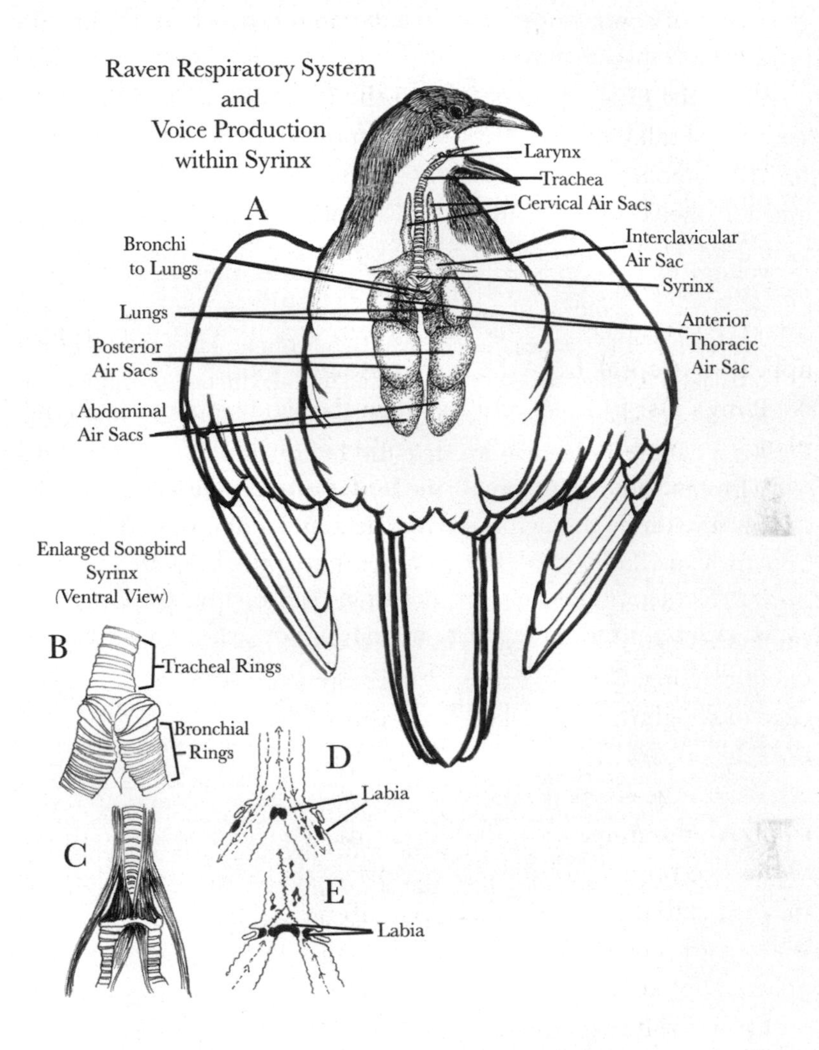

The ability of crows to sing and speak is a function of the songbird's remarkable syrinx. (A) The birds' respiratory system features a trachea that splits into two bronchi attached to a pair of lungs just as in humans. But bird lungs are relatively rigid structures finely dissected by ever smaller branching air tubes, rather than the inflatable balloonlike lungs we possess. Inhaled air passes into a bird's lungs and on through to the set of posterior air sacs, which inflate like our lungs in response to a big breath. When the bird exhales, the inhaled air moves forward into the lungs and passes oxygen to the blood. On the next inhale as oxygen-rich air comes into the lungs and back to the posterior air sacs, the first breath moves from the lungs to the anterior air sacs. With the next exhale the breath finally leaves the bird, having given up all its oxygen to fuel the demands of an active physical life. This unique two-breath system ensures that a bird's lungs are always in contact with oxygen-rich air and that the entire volume of a breath is exhaled, thereby maximizing the addition of oxygen and removal of carbon dioxide. (B) The syrinx sits at the point where the trachea divides into two bronchi. Here several of the cartilaginous rings that support the flexible air ducts fuse and harden into a bonelike resonating chamber. (C) A songbird, like the raven, has a complex musculature that the brain controls to deform the syrinx and produce sound. As the bird inhales or exhales silently (D), air flows freely through the syrinx and into and out of the air sacs and lungs as described above. When quiet, the labia are staggered kitty-corner fashion, one on the inner and one on the outer side of each bronchus, so as not to impede airflow when the bird is not vocalizing. But if the bird is vocalizing (E), the specialized syringeal muscles contract, shoving the syrinx mouthward and compressing the lower trachea. This forces the outer wall of the bronchial portion of the syrinx to move upward and inward and align the crow's lips. In this configuration exhaled air vibrates the lips producing the notes and noises a bird uses to sing and talk. Because each bronchus has a set of noise-producing lips, the songbird can produce notes from two instruments and then blend the song as it leaves the mouth.

Crows' ability to speak, while enabled by their singing apparatus, depends on their brainpower. Their large forebrain and a specialized loop between vocal centers in the forebrain and thalamus are key features that shape what a crow hears into what a crow says. Let's follow the voice of a human into a crow's brain and see it transform from the familiar, ordinary speech of a person to the surprising utterance of a bird.

One can say that when a crow hears a person speak, its hair stands on end. Yes, birds do have hair—deep within their inner ears. As in mammals, it is the inner-ear hair cells that first translate the mechanical pressure waves of sound into electrical signals rushing toward the brain for interpretation. You may never have seen a bird's ear; instead of fleshy external structures like we have, there are small holes under the feathers on each side of the head where sound enters. Sound waves travel down this external ear to the eardrum, which vibrates and rocks the liquid that fills the middle ear. This sloshing causes the single, rod-shaped ear bone (we have three, birds only one) to rap against the membrane of the inner ear.

The inner ear of a bird is a wild concoction of tubes and loops reminiscent of a gyroscope. In fact, it is gyroscopic—and keeps the bird's nervous system informed about the bird's orientation so that it can maintain its balance and equilibrium. One part of the inner ear, the cochlea, houses the sensitive hair cells that enable birds' keen sense of hearing. You will find a diagram of the cochlea in the Appendix, pages 224–225. A crow's hair cells are dense—ten times as dense as ours—and so diverse that no two may be the same. There are tall ones, short ones, some facing in every direction, bundled ones to amplify soft sounds, some with nerve inputs, and some with both inputs and outputs. This diversity and density enable birds to hear high-pitched squeaks and soft rustles in leaves, under snow, and beneath the bark and soil. Our hearing range, as well as that of most mammals, is comparable to that of birds, and it allows us to hear and recognize their songs. Yet we humans cannot hear sounds as faint or distinguish as quick a beat as can a bird. A bird's hearing is also more resilient. Unlike in mammals, hair cells in a bird's ear damaged by loud sound or toxins regenerate. Old birds don't need hearing aids.

The nerves from the crow's hair cells usher sound toward its brain over circuitry that evolved over 300 million years ago. This same auditory circuit exists in crocodiles and humans. It is a heritage from our earliest egg-laying ancestors. Sound signals flash from hair cell to sensory nerves in the brain stem of all reptiles, birds, and mammals. From there, relays occur in the midbrain, thalamus, and forebrain. Different structures of common heritage transmit the message of sound in reptiles, birds, and mammals. In birds and mammals, sounds processed in the forebrain provide neural feedback to the auditory centers in the thalamus and midbrain. Birds and people have specialized areas for recognizing familiar sounds. Wernicke's area in our brains allows us to rapidly recognize words, natural noise, and music. Damage to this part of the cerebrum causes people to be unable to recognize sounds they know. Similarly, songbirds cannot remember familiar tunes if their cerebral auditory centers are damaged.

The hearing pathways of all birds and mammals evolved long ago, but the ability to control and modify vocal output by learning is a relatively recent innovation that only songbirds and six other groups of animals have developed. Humans and other primates, bats, cetaceans, elephants, parrots, and hummingbirds are the other vocal learners. Among mammals, bats or elephants were likely the first vocal learners when they evolved over 50 million years ago. Parrots evolved before hummingbirds, which evolved before songbirds, so it is likely that avian song learning first occurred in parrots, some 85 million years ago. Syrinx-equipped songbirds evolved 70–80 million years ago and bestowed song-learning brains upon ancestral crows some 50 million years ago. Birds, bats, and other primates had a significant head start over humans who evolved only 6 million years ago.

The song- and speech-learning systems of songbirds and people are well known and, despite their independent evolutionary paths, remarkably similar. Both involve neural interaction with their auditory systems: in people this includes extensive involvement of multiple thought centers within our forebrains; in birds multiple forebrain regions are likewise involved in song learning and song production. The same gene (FoxP2) stimulates nerve development and connectivity important to language and song learning. Auditory memories of songs and words circulate through both bird and human brains on specialized loops connecting the thalamus and the forebrain. Humans and birds both excel at remembering auditory information. Sleep is critical to song learning and language acquisition; REM sleep reinforces, while slow-wave sleep consolidates learned vocabularies.

When Macaw heard Tony greet him each day, that sound rippled from his ear up his eighth cranial nerve and into his forebrain and eventually became a memory. Macaw could recall perfectly and remember the greeting for future use and elaboration. We know this because hundreds of experiments, surgeries, observations, brain scans, and recordings from single nerve cells within the brains of songbirds have been conducted by scientists over the last half century. Most of this research involves typical singers, like sparrows, that listen to and learn from other members of their own species. Learning the correct song is important if a sparrow is to later secure a territory and attract a mate.

A young sparrow is hatched into a noisy world where it selectively learns the songs typically only his father or neighbors sing. The young hear and memorize these songs and practice and perfect them by comparing what they hear to what they remember during a sensitive learning period. A fully formed, "crystallized" song emerges the following breeding season and remains fixed for life in most songbird species. Some birds, like crows, ravens, magpies, catbirds, and mockingbirds add to their collection of songs and modify them throughout their lives. They do not have a fixed critical learning period but continue to soak up new sounds from their environment, remember them, and reproduce them. But the ways in which birds learn new sounds—whether they be the sounds of crickets, screams of hawks, or mutterings of people—involve the same neural pathways.

The development of Macaw's vocabulary was already underway as he listened to Tony. Up until a few days before fledging, he was raised by wild ravens and knew many of the typical raven sounds. Tony's words, now coded as a particular pattern of nerve-cell electrical activity in the thalamus, converged on the auditory regions of Macaw's forebrain. Tony's voice was received, filtered from background sounds, and processed by three complexly interconnected regions likely important to memory formation and storage (please see the Appendix, pages 224–225, for a depiction of the bird's hearing pathway). Neurons in these regions respond to the familiar songs and calls of an individual's own species. In the case of talking crows, ravens, and magpies, this sensitivity apparently extends to the speech of their close human associates. Sounds that stimulate these regions may provoke a raven to talk or to remember the voice.

Tony conversing with Macaw.

The electrical rendition of Tony's voice in Macaw's brain—which neurons were activated, when they were activated, how long they were active, and how they influenced other neurons—passed from the sound reception area of the forebrain into the sound-production and song-learning circuit. A first step in this circuit is a region of the hyperpallium known as the HVC (we diagram the song-learning circuit in the Appendix, pages 226–227). Essential to a bird's ability to learn speech, the HVC is one of seven interconnected, cerebral regions in the songbird's brain that are analogous to the regions in our brains, like Broca's area, that control our vocal learning and talking. When the equivalent region of a parrot's brain is damaged, the parrot cannot learn to speak.

The HVC contains at least two types of neurons that act either to motivate an immediate vocalization or update the song-learning process. To talk, Macaw's brain codes electrical signals destined to provoke the muscles coordinating breathing and airflow in the syrinx. Some neurons in the HVC have mirrorlike properties, firing most vigorously in response to sounds that are being imitated. This would allow Macaw to instantly transcribe Tony's voice into the electrical signals required to appropriately flex his syrinx and breathe forth a human word. While accurate, the first transcriptions may not be perfect, so Macaw's brain may also loop these codes through other regions in the song-learning circuit so that they can be reconsidered and revised. In Macaw's striatum (Area X in our Appendix diagram, page 226) successive renditions of the electrical code that mirror's Tony's voice can be compared and improved. This is how a songbird reshapes its song to imitate an important tutor or add a neighbor's sexy flourish. Macaw might polish his pronunciation, or expand his vocabulary with a new phrase. This integration of new sounds into an existing song collection involves extensive feedback between what a bird hears and what it already knows. Old memories are bombarded by new sounds, some of which work their way into the repertoire, others of which are forgotten.

The neurotransmitter dopamine plays an important role in songbird vocal learning. This neurotransmitter (chemically similar to histamine and adrenaline) is a potent motivator that causes animals to seek rewards and helps them learn complex associations. When an important food, a valued partner, or a challenging rival is encountered, dopamine released into the bird's striatum motivates eating, singing, or courting. Dopamine does this in part by strengthening active synapses, increasing the ease with which additional stimulation promotes their firing. During song learning, dopamine may motivate the learner, indicate the quality of its song, and reinforce the accuracy of imitating another sound.

As Macaw's representation of Tony's accent improves, for example, the neurons in his striatum increase their stimulation of other neurons in the midbrain, which increase their feedback of dopamine to the striatum and other forebrain areas. By keeping dopamine flowing, the synapses between neurons in the song-learning region of the striatum and those providing new and remembered sounds from the HVC and elsewhere may be strengthened, allowing continued refinement of a vocal memory before it is used to guide speech or song. Forebrain and thalamus in essence collaborate, through reciprocal chemical feedback loops, to fashion the bird's song (as detailed in the Appendix, pages 226–227).

The neurons in Macaw's striatum likely increased their rate of firing as his own rendition of Tony's voice came to sound more and more like Tony. By comparing the voice in his head to the voice streaming in his ears, Macaw's brain adjusted its chemistry to perfect his voice. The motivational presence of dopamine may explain why socially important sounds, like the voice of a partner—Tony to Macaw—may be priorities to remember, as well as sounds that are heard in fearful or pleasurable situations. This could explain why the crow that spoke to Professor Lorenz accurately remembered the voice of his trapper and why French and American crows faithfully reproduced the phrases of human mothers and dog owners.

Other important chemicals helped shape Macaw's voice. The proteins produced by a gene shared by humans and birds, FoxP2, may have guided the elaboration of Macaw's repertoire in his striatum. Just as these proteins aid human speech learning, they regulate the flow of information from Macaw's striatum to other brain regions important to his language acquisition. In this way, FoxP2 is an editor of sorts, allowing some memories to be consolidated into an existing repertoire and denying others. When the proteins produced by FoxP2 genes are abundant in the striatum, the information required to produce a memorized sound loops through the thalamus and back to another song nucleus in the forebrain (in the Appendix diagram, page 226, this is "LMAN"). This area is the bird's center of vocal creativity; when it is damaged, a young songbird sings a consistent song and does not experiment with new sounds. Information that is not looped is deleted, removed from consideration for inclusion into an expanding range of vocal expressions. Macaw's memory of Tony's voice was in an area of his forebrain that was flooded with dopamine and the products from FoxP2 genes. And because of this, a human phrase became part of a raven's vocabulary.

Tony's voice is now almost ready to be spoken by Macaw. Macaw has memorized a phrase, rehearsed and practiced it in his head, and compared his memory of it to Tony's daily greetings. The memory of Tony's voice has cycled many times along the songbird's forebrain learning loop, allowing Macaw to perfect his imitation. From here the memory is passed to the motor centers of the forebrain where it guides the voice to match what has been learned.

Macaw processes and remembers Tony's voice noisily during the day and at night, silently. He is cogitating, thinking. And dreaming. In REM sleep, the nerve cells in the forebrain fire just as they do to produce a sound. But rather than making a sound during sleep, a corollary discharge from the forebrain to the thalamus routes back to the song-learning pathway. Inactivation of Macaw's muscle pathways during REM sleep ensures that he doesn't talk in his sleep. Yet this auditory replay allows a sleeping raven to perfect his accent. Sometime, perhaps tomorrow when he sees Tony or hears him speak, the electrical code of Tony's voice will be released from Macaw's forebrain, sail through relays in his midbrain and hindbrain to the nerves that control breathing and the syrinx, and Macaw will utter a perfectly formed, *"Hello, Macaw!"*

All the crows, ravens, and magpies that learned to speak did so with mental practice. Their learning required extensive revisiting and refining of memories. This was enabled by the coordinated actions of distinct regions within their brains, just as within our own. Not all vocalizations spoken by a crow require such cognition, however. Some of their emergency signals, shrieks, and cries are fully formed at birth. But learning to imitate human speech is certainly an advanced cognitive trick that points to an important parallel between crows and people—our shared reliance on lifelong learning to shape our vocabularies. Even so, this learning capacity seems only part of what talking crows are telling us. When the birds call to French poodles or children, these talking crows seem to understand what they are saying and are using their sharp tongues for personal gain. Talking crows are likely former pets, but talking crows are thinking crows.

A crow might understand what it is saying as it is *imitating* human speech. To scholars of animal behavior, imitation is the intentional copying of an improbable, novel act. Imitation implies that the crow understands the purpose of human language, the goals of a talking person. Primates and people are imitators, but until recently birds were considered mimics. Mimicry is simple, mindless repetition. Mimes like fireflies flash their lights in particular sequences to attract mates. Some flash the sequence of another species, and when a horny fly homes in on the sexy light, the flasher becomes a predator.

Imitation requires thought, planning, and coordination by the brain's cognitive centers. Mimicry may be reflexive, dependent on specific immutable nerve circuits, and need not involve the forebrain. Some learning of human phrases by birds may indeed be mimetic. Harry the magpie's incessant "*no, no, no*" to any sort of query seems like mimicry, even if it is a learned response. But careful study of parrots and consideration of our crows suggest that some birds also are true imitators.

Alex, an African grey parrot made famous by the research of Professor Irene Pepperberg, used and understood more than one hundred English words to demonstrate his command of abstract concepts like shape, color, texture, and quantity. His use of language far exceeded what we know crows to be capable of and rivals the abilities of signing apes. The detailed observations made by Dr. Pepperberg on Alex during their nearly three decades together show that Alex imitated, rather than mimicked human speech. The most compelling evidence in favor of imitation comes from the way Alex learned a new word. At the age of twenty-seven, Alex had never learned to say *spool*. But he watched and listened to another parrot, Arthur, learn to say *spool* as he played with wooden or plastic bobbins. Arthur followed a regular course of enunciation, starting with *ooo* and graduating to a fully formed *spool*. Alex, however, did not start from scratch. Alex combined previously known sounds to create the new, arbitrary label for a toy. He mouthed *s,* something he knew from alphabet training. And he mouthed *wool,* another object he could identify. For a year he said *s*-pause-*wool* when he interacted with bobbins. Then, Alex spontaneously produced a perfectly enunciated *spool*. Alex did not simply memorize and repeat a label but crafted it from shreds of knowledge he already possessed. This restructuring of his vocal repertoire occurred in his forebrain's song-learning network, probably after a sound sleep. Alex understood arbitrary labels, learned to use them in reference to important items in his world, and demonstrated his understanding that labels are built from individual units that can be intentionally recombined. This intentional recombination is evidence of imitation, not rote mimicry.

We have no reason to doubt the ability of corvids to imitate, but there is no verification that crows intentionally recombine sounds to produce new words, or rearrange words into novel phrases or sentences. Parrots and crows learn sounds, including human words, by tapping neural regions and circuits that share a recent, common ancestor. The dedicated, personal, lifelong research Pepperberg did with Alex has been done in only a handful of situations with other species. Crows would make a great subject for this kind of intense one-on-one investigation.

African grey parrot.

Without tracking the acquisition of speech by crows, we cannot know if words and phrases are learned in their entirety and successfully uttered upon first try, or built piecemeal as suggested by Alex's behavior. We think it likely that eavesdropping crows learn phrases or words as entire units and imitate them. We infer this from the fact that these speaking birds uttered words in improbable, novel, and completely appropriate settings. They appear to have clear goals, and their musings help them attain those goals.

Crows may recombine known words or create new ones, but all those we know have limited vocabularies and have not been trained to associate individual words or sounds with rewards. Alex knew individual words and letter sounds because his training was designed to teach him these units. His life revolved around watching people interact with an object and repeatedly say its name. To obtain the item, which was fun to play with, Alex needed to say the word or voice the letter. Talking crows are not trained this way. The crow's energetic character has inclined it to coevolve with humans through a more distant although exploitive association, whereas the parrot's less rambunctious behavior has routinely made it a favored pet where it is in far more intimate and constant contact with people.

Crows are nevertheless keen observers and listeners. They distill the elements of human speech from ongoing conversations. Just as we have difficulty hearing word breaks in a foreign language, so, too, would crows have difficulty recognizing the arbitrary subunits of human speech. This does not mean they do not understand that speech is composed of subunits, but simply that they may recognize different subunits than do humans. Upbringing, exposure, and the mechanics of hearing are likely reasons why talking crows sound like they are mimicking rather than imitating.

Intentional speech by free-flying crows gives us a clear window into their psyche—they know that some utterances produce predictable, self-rewarding outcomes. Roah was rewarded by attracting Professor Lorenz. Spike the magpie may have been rewarded by a surprised reaction from Ms. Woolfson. Controlling another species and stimulating mayhem repeatedly rewarded the Missoula campus crow. Calling dogs and holding them at bay until people arrive is not a simple emotional response to a pressing and current situation. This vocal mastery includes planning for a future event. Whether the words and whistles are mimicked or imitated, their use is goal oriented and anticipatory—intelligent.

Corvids' use of phrases, posing appropriate answers to questions, and distinguishing among pronouns suggests that these birds are capable of thinking and planning before they speak and that they actually appear to understand basic aspects of our language. Crows may even understand and interpret the tone of human speech far earlier than they understand specific words. Most linguists, philosophers, and animal behaviorists would say that crows and other songbirds do not have a communication system organized like human language. Human language differs from other animal communication systems, even those where animals speak or sign their intentions, by the degree to which humans use rules to combine a few discrete, arbitrary sounds into a nearly limitless and readily understood group of novel and complex words, phrases, and sentences. Alex taught us that birds are capable of such recombination, but his abilities pale in comparison to those of a young child. And, beyond the structured merging of two or three calls into compound messages by some birds, this skill is unknown in any wild animal.

The natural and imitated language of crows shares a key feature with human language—the use of arbitrary symbols (words) to represent concepts, objects, and relations. Arbitrary calls are not reflexive like cries or screams, nor are they iconic like the mimicked calls of predators. Macaw represented Tony with the arbitrary phrase *Hello, Macaw*. The Missoula crow took advantage of the dog-owner relationship encoded in *Here Boy!* Spike, the magpie, represented a concept by answering, *What?* These words were learned to have a particular meaning, and unlike onomatopoetic words, the sounds themselves have no inherent meaning. Different words could have been learned to mean the same thing, just as in English and Spanish "good day" and "buenas dias" are learned to mean the same thing. Our words have no inherent meaning to a corvid; they are arbitrary, but the natural communication system of these birds also involves arbitrary symbols (calls) that refer to specific objects and actions in their world. Ravens *cluck* like hens when they detect danger near their nest, but crows use harsh *caws* to indicate danger and recruit help. Neither sound is naturally associated with danger. In contrast, crows also imitate the calls of some predators, like hawks. When crows sound like a hawk to signal danger or alarm, they are doing so with a logical, not arbitrary, noise. Hungry ravens that see but cannot obtain meat give ethereal calls that sound like *haaaa, haaaa*. More mundane fare, like pig chow, rarely sets a raven to *haaa*-ing. *Haaa* calls refer only to meat, and when ravens hear the call, they anticipate feeding. *Haaa*s are arbitrary signals of hunger, while stomach growls or imitations of a wounded animal's cries would be less so.

It is a long way from the use of arbitrary, referential signals to a rule-based, combinatorial communication system like the human language. But language discovery in a wild animal will not be surprising. As we scientists become more sophisticated in matching the response of animals to nuances in their vocalizations, we may discover language in cognitive birds and mammals that rivals our own. Crows *caw*, for example, in complex and regular patterns and emphasize messages by lengthening the sequence of *caws*. But could there be more to this? Their sequences of calls could be true combinations with emergent meaning, like our phrases and sentences, although our current ability to measure call sequences suggests this is not the case. The future may bring finer resolution.

Even if true language is discovered in an animal other than humans, we still suspect human language will remain unique in its ability to represent future possibilities and past occurrences and our propensity to talk about them. As far as we know, animals talk about the here and now. What they say is motivated by an immediate need, and the utterance is intended to fill that need. Birds sing to attract mates and defend territories now. Ravens *trill* to signal threat, *cluck* to signal danger, *beg* for mercy, and request food with a British accent now. Future events may be planned and coordinated with vocal signals, but even dog-calling crows are not known to gather and gossip. There is no known episode of crows sharing sagas. King Solomon's ring will require a lot of rubbing before we can eavesdrop on those discussions, should they exist.

Natural selection is intolerant of idle verbosity. A crow that talks about possibilities may quickly become a hawk's dinner. Only if speculation about the future or reconsideration of the past has survival and reproductive benefits would we expect to see it evolve in species other than ourselves. That seems unlikely, although selection does sometimes favor verbose birds. Individual crows build extensive repertoires of natural and imitated sounds that they use to identify and attract important social partners. Mates, who remain together for life, recognize each other's voices, participate in well-structured duets, and may label each other just as Macaw labeled Tony.

These social needs, while not immediately pressing like the need to get food or warn of danger, do affect a crow's ability to survive and reproduce. And they favor individual identity in voice and individual variability in the sounds social animals make. Wider social circles may favor particular arrangements of calls into characteristic songs. Crows that live together in flocks, for instance, imitate one another's songs. These songs work like passwords. When a crow joins a new group and does not sing the correct song, it is attacked. As it learns the group song, aggression decreases and it is able to enjoy the benefits of group life. Sociality begets verbosity and vice versa, as dialects in American English set people apart and a common accent unites.

Talking crows reveal a part of their cognitive lives. To talk, crows must be able to form and replay memories. They confront the immediate with memory of the past. They dream. While we don't claim that speaking crows really grasp the complexity of human language, they use our words to get what they want, which is remarkable. That a crow will learn and use a human trick reinforces the depth to which our species are intertwined. Crows manipulate, deceive, play, and converse with other species. They anticipate rewards and, to reap them, devise and carry out plans. When we overhear crows singing softly to themselves, we wonder if they derive pleasure simply by listening to the sounds they can make. So much of what we hear from crows or ravens is inexplicable. They ring like bells, drip like water, and have precise rhythm. They sing alone or in great symphonies. Some of their noise could be music.

4

Delinquency

Hitchcock, the raven who stole windshield wipers
from visitors to the North Cascades.

CATHI JONES, a U.S. National Park Service biologist, was anxious, maybe even unconvinced herself, as she explained the situation confronting visitors at the Newhalem campground set deep in Washington State's Cascade Mountains. She told us that many of the cars and trucks parked in the campground were having their rubber wiper blades systematically destroyed. Sometimes, even the stout metal of the wipers was twisted and mangled, but more typically the rubber was surgically removed and the deed hardly noticeable. The park officials feared that drivers, unaware their wipers had been pilfered, might be caught defenseless in the all-too-frequent mountain rain and snow storms. The biologist had identified the perpetrator, but because it was a raven, no ordinary arrest could be made. Besides, how was she going to catch him? Exasperated, she got in touch with us and asked if we could help. We agreed and reminded her that he was protected from incarceration by federal law, the United States' Migratory Bird Treaty Act.

Park employees named the windshield-wiper perp "Hitchcock." Warning posters with his picture and *modus operandi* were all over the camping and picnic areas. We got our first glimpse of the villain, flying defiantly with his lifelong mate, on August 4, 2006. A territorial common raven living at the interface of wilderness and rural America, Hitchcock had been emboldened by a diet rich in salmon, roadkill, native fruits, and small animals that was supplemented by tourists' sandwiches, bags of chips, and grilled delicacies. The ebony bird ran the joint. Not only did he steal windshield wipers, but he was implicated in several break-ins at nearby private cabins—tearing screens from doors and windows and removing nearly all the rubber molding from a large camping vehicle. Adding insult to injury, he regularly whitewashed the fancy glass, rock, and woodwork exterior of the Newhalem Visitor Center with a messy calling card. He scratched and pecked at the windows each day before the first shift of rangers could shoo him away.

We had to travel north and see for ourselves. What motivated the mischief? Were Hitchcock and his mate the only good ravens to go bad? How could we put an end to this "crime spree"?

John and his daughter, Danika, mulled over these questions as they sat in their truck and hoped Hitchcock would find a booby-trapped loaf of bread dropped before dawn in the rustic parking lot. We had carefully hidden a remotely triggered net in the ferns beside the bread and now sat, finger on trigger, ready to catch a thief. We did not have to wait long. It was barely light enough to see when we heard the deep *Quork! Quork!* A raven pair was chatting and heading toward the visitor center. The birds saw the bread, but nothing of the trap, and glided down to the asphalt. They were easily twice the size of our neighborhood crows, carbon colored, demonstrative, and hungry. At trapside, John could hardly breathe as they walked like generals to claim the morning's bounty. As both birds beaked a slice of bread, he triggered the net and with a resounding *boom!* the ravens were snared. A second later Danika and John were untangling the two surprised ravens. Our plan had worked perfectly. Next we restrained the birds, measured them, and decorated them with leg bands (rings) of individualistic color. The bands would allow park officials to determine if this was indeed the delinquent raven and his mate. Soon we would know if there were accomplices. It usually takes about ten minutes to measure and mark each bird, but we worked a little slower, hoping to aversively condition the ravens in the process. We banded the birds by pinning them against the windshield and atop the wipers of our truck. Here the ravens could get a clear view of these accessories that we believed they had taken such an interest in. We hoped their brains would get the message: associate windshield wipers with this fearful encounter.

We aimed to rehabilitate the thieves so that more drastic methods could be avoided. Delinquent ravens and crows are often simply shot, but this is rarely effective, and is illegal. New birds move into formerly occupied territories and the mischief continues. By teaching a pair how to behave, we hoped these resident birds could remain and the problems subside.

"Aversive conditioning," the method we used to reeducate the ravens, is not a new idea. It builds on experiments that the Russian scientist Ivan Pavlov pioneered with dogs in the 1920s. Pavlov, and many scientists since then, demonstrated that animals quickly form associations between good and bad events and features of the environment that predict such events; bells that precede food, lights that precede shock, for example. To make these associations requires a bit of neural rewiring. In many vertebrates, including people, the rewiring involves stronger integration between neurons conveying stimulation from the sense organs and those in the amygdala that acquire and modulate fearful responses. Learning associations in general relies in part on increasing the number or sensitivity of synapses between interconnected neurons. Learning fearful associations specifically seems to depend on reinforcing the synapses between the neurons that are simultaneously firing in response to environmental cues and those in the amygdala that are firing in response to danger.

Here is how we hoped it might work on Hitchcock and his mate. The sights, sounds, and pressures of the trapping event immediately triggered the production of stress hormones in the ravens' adrenal glands. Adrenaline (known more formally as epinephrine) pushed their muscles to spring away from danger. This chemical might still have the birds ready to spring away to freedom if we loosened our grip. But another hormone from the adrenal gland, corticosterone, works over a longer period of time to enhance memory formation in the integrative amygdala. Corticosterone ferried by the blood diffuses into the brainstem and hypothalamus, triggering the release of norepinephrine into the amygdala. This chemical increases the attention paid by the bird to sensory information about typically harmless aspects of the environment and information about a very specific danger.

When we fired the net, the brains of our ravens received information about the trapping scene that was directly frightening—a nearby gunshot (discharging a blank rifle cartridge supplies the force needed to propel the net over the birds), physical restraint, and information that until that very moment was harmless—our faces, the pile of ferns, the white bread, the truck windshield. Electrical signals deciphered and routed by the eyes and ears stimulated a chain reaction of neurons in the ravens' thalamus, forebrain sensory and integrative centers, and the amygdala, where electricity collided with hormones. Epinephrine and corticosterone bind to neurons, increasing the ease with which active amygdala neurons form lasting synapses with those in other regions of the brain.

This strengthening of synapses with neurons privy to context and history from throughout the forebrain can cause a fearful raven to associate normally harmless sights and sounds with danger (we provide a diagram of this in the Appendix, pages 228–229). If Hitchcock and his mate's neurons that were activated by the sight of the windshield wipers beneath them, the nearby visitors center, or the general location where we stood, connected with neurons in their amygdala that were firing because of the very real danger of our restraint, then these synapses would be strengthened. As a result, normally unimportant features of their world would become more strongly linked to take on new, fearful meaning and become objects of avoidance.

When we released Hitchcock and his mate, we hoped that their new lessons would be among the most important of the day. If so, they would have a nightmarish sleep as the linkages between visions of windshields, memory of the trapping location, and fear of restraint would be replayed, strengthened, and consolidated during REM sleep. Just as Macaw dreamed of Tony's voice, Hitchcock should dream of being captured. The next time the ravens approached the visitor center, the sight of a sandwich, loaf of bread, a mound of ferns, or the windshield of a car would fire the interconnected neurons in their amygdala, hippocampus, and forebrain sensory centers that together command their muscles to flee. We reasoned that fear and nightmares were better than death.

Over the next year, we were able to confirm that Hitchcock and his mate were alone in the crime. As the resident pair, they had exclusive use of the visitor center and the nearby campgrounds, as well as the more distant private residences. They also became model citizens, in part due to the strengthened synapses within their brains, and in part due to the actions of park officials.

After we trapped them, the ravens rarely landed in the parking lot, but for a short time they continued to foul the visitor center windows. We tried to trap them again at the windows, but just the sight of fern fronds piled over the net sent them flying. By keeping potted ferns around the windows and mounting posters of raven predators inside the windows, rangers were able to keep the birds at bay. Hitchcock and his mate had acquired a new fear of potted plants, specifically, and the visitor center, generally. The incidence of wiper stealing dropped gradually after trapping. Perhaps our aversive conditioning worked, but park rangers also made sure that fewer wipers were available to tempt ravens. Visitors entering Newhalem were given plastic pipes to slide over their wipers while they parked. Now Hitchcock not only had less motivation to steal, but the opportunities to do so had diminished as well.

With Hitchcock reformed, we could think more about what sent him down a path of crime in the first place. We considered five hypotheses. Perhaps Hitchcock was an anomaly or just another pet with bad habits. Lawrence Kilham, an ornithologist from New England, reported that his pet raven routinely mangled windshield wipers. Possibly revenge was involved, and this was Hitchcock's means of harassing a perceived enemy after one of his offspring had been run over by a car or perhaps a car camper had threatened him. Hitchcock could have been enforcing his territory ownership. Flying over a parked car or approaching a window, he might see his reflection and think it was another bird on his turf. A black wiper might fool an enraged bird amped on testosterone into thinking it was the wing of a rival. Alternatively, the resourceful ravens may have seen food in the car, the center, or in the cabins. Or maybe the wipers themselves were somehow useful to the birds, although a cache of rubber wipers has yet to be found. Finally, maybe Hitchcock enjoyed or benefitted in some less direct way from mischief.

Anomaly or revenge seems an unlikely explanation. Vengeful corvids are usually more direct. People who purposefully or incidentally endanger a young crow quickly end up in trouble: they receive a tongue lashing from a scolding parent or are easy marks for aggressive dives and occasionally wallops. Stealing seems too subtle and sophisticated for a raven's revenge.

A fascination for wipers is not anomalous. A thousand miles south in Yosemite National Park, ravens take wipers. This habit is practiced to the North, in British Columbia's Cypress Provincial Park, and in Kodiak, Alaska. Even east to New Brunswick, Canada, ravens take wipers. In Wageningen, Netherlands, some crows have picked up the habit of swiping wipers. There is something about rubber. Ravens even peel off expensive, rubbery radar-absorbent material on structures at the Naval Air Weapons Station China Lake, deep in the Mojave Desert. Ravens are not the only rubber addicts: the black parrot, or Kea of New Zealand, filches wipers and, if undeterred, will completely disable a parked car by pulling out all the rubber lining around the windshield.

Aspects of Hitchcock's rampage are consistent with territorial defense. Recognizing one's own reflection in a mirrored surface is uncommon in the animal kingdom, and ravens, at least those with little exposure to mirrors, threaten their reflections. They act as if another bird—one intent on attacking them—is challenging their homeland. Hitchcock may have been fighting an unrelenting, reflective foe. But this is not the only reason he was at the visitor center windows. When rangers covered the glass with dark paper to block reflection, Hitchcock did not acquiesce. He tore the paper off and went back to throwing himself at the windows. He wanted in. Inside the center were images of ravens: Native American–style murals that were larger than life. There were also sounds. An interpretive film about the region featured a soundtrack rich in raven calls. Certainly Hitchcock could hear the territorial calls of the raven on the video; he may have been tricked into thinking there was a raven living in the building. Such an intruder was all the more challenging because on the north side of the center, just under the protective eaves, was an old raven's nest. The nest may have been built by Hitchcock and his mate in the years before the center opened to the public. This was sacred ground: the center of his territory, the bedroom. Hearing an intruder in that setting, every day, would make any animal snap.

Territoriality may explain Hitchcock's rude behavior at the visitor center and may also account for his savaging of nearby windshield wipers. His reflection may have lured Hitchcock to explore windshields. Frustrated animals often release pent-up energy by attacking the nearest object. Gulls, for example, often pull grass when facing off with neighboring rivals rather than crossing into another's territory for beak-to-beak combat. This "displacement behavior" could apply to Hitchcock, who, unable to get at his reflective rival, pulled at whatever was within reach. Rubber wipers, rubber seals, and bits of paint or trim would be obvious targets. That most of the missing wipers were from long-parked cars (typically several days) supports this idea; rivals that won't leave may be especially frustrating.

The behavior of a cognitive animal like a raven rarely has a simple explanation. Territoriality and frustration may contribute to Hitchcock's tirade, but we also suspect food was a partial motivator. Hitchcock must have repeatedly witnessed people eating in and around their parked cars. Maybe someone even fed him from or on a car, reinforcing an association between automobiles and food. Many of the damaged cars in the North Cascades held no food, but if Hitchcock was motivated to search for food and found none in a usually reliable spot, like a child who can't reach the cookie jar and throws a fit, he may have taken out his frustration on the wipers or windows. When one considers the relative pliability of the rubber, the raven, from past foraging practices of skinning and tearing spawned-out salmon or the hide on a roadkill, may have considered this a soft point of entry to reach the contents of the car where previous campground experience proved it to be a source of food.

Finding something to eat in nearly any environment, from our largest cities to the high wilderness of the Himalayas, is a key to corvine existence. Selecting, exploiting, and anticipating new foods in strange places is accomplished in much the same way as is learning about new dangers. While the amygdala is an important junction box in the brain that underwrites aversive conditioning, the region where neurons stimulated by eating influence the synapses of those carrying sensory information from the environment is not precisely known. Junctions may occur in many parts of the avian forebrain, such as the hippocampus, or the striatum, where sensory information and spatial information converge. Neurons carrying electricity from taste and smell sensors synapse with neurons firing at the places where food is found as well as those only mildly stimulated by other sights, smells, sounds, and touches—potential cues to new foods. These latter synapses may be strengthened to form predictive associations between environmental cues and a good meal. A contextually and emotionally relevant association is forged by synapses among many brain regions. This can be thought of as a connected circuit, the neural representation of a learned association, among regions of the brain that convey history, emotion, sensory input, and muscular output. Particular cars, people, textures, odors, places, bags from fast-food restaurants, or times of the day may thus become cues to a corvid's feast.

Dopamine may be especially important in guiding the dining decisions of corvids. Monkeys, when finding food, experience an increase in the firing of neurons in the midbrain that deliver dopamine to a variety of locations in the forebrain. The presence of dopamine increases the formation of functional connections among brain regions, and the relative rate at which dopamine neurons fire may be a mechanism for remembering environmental cues that predict foraging success. In mammals, midbrain dopamine-producing neurons refine their firing patterns after they make associations between food rewards and environmental cues. Rather than being triggered by the sight of food, they fire when reliable cues, such as the track or burrow of a tasty insect, are spotted. Our crow panhandling at the ferry landing anticipated a good meal when the cars lined up because dopamine produced in the midbrain was delivered to its forebrain.

Seeking food, while innovative, may be at the root of a long list of corvid behaviors some find offensive: killing weak lambs in Australia; unzipping backpacks and saddlebags in the wilderness; stealing

In Japan, jungle crows steal candles from shrines, occasionally setting nearby fields ablaze. The tallow used to make the candles may be attractive to hungry crows.

fatty soap from children and tallow candles from shrines in Japan; eating endangered beach-nesting piping plovers, desert tortoises, and greater sage grouse in the western United States; peeling back freshly laid sod in California to get at grubs; popping toads for a bit of fresh liver pâté in Germany; and slaughtering 141 eared grebes who mistakenly landed on a frozen lake in Yellowstone National Park. The grebes were dispatched by four ravens, who ate their fill and cached the rest. We suspect their dopamine levels were off the typical feeding bird's chart.

While their skill at innovation has given crows and ravens a bad reputation, their problem-solving abilities are inarguably impressive. In Lake Washington, just east of Seattle, crows have taken up fishing. They hover above the water's surface and with their beaks pluck out small sticklebacks and sockeye salmon fry. In Clear Lake, California, crows plunge from the sky into the water like osprey to try to snag fish, but mostly they just get wet. Ravens in Grants Cove, Alaska, swim as much as they fish: they jump right in the cold ocean and swim around fishing boats like surface-fishing cormorants looking for stunned prey. On land, ravens are more patient. One was seen crushing tunnels made by voles under the snow, then waiting and watching for the small rodents from a nearby perch. This unique hunting strategy demonstrates imagination—the ability to visualize an out-of-sight object, as well as planning and anticipation. Near our homes in Washington, another raven was observed persistently jumping from one side of a sheet of plywood to the other, rocking it back and forth until a mouse, which the bird had watched retreat there, was flushed. Once out from cover, it became the raven's meal. Out of sight is far from out of a corvid's mind.

Sometimes innovation requires tools. Crows relish roadkill and use our cars to their advantage. We have observed half a dozen crows working in a loose formation continuously dive over a flock of rock doves that were coursing back and forth over the interstate near San Francisco. They forced the doves lower and lower until they'd herded the flock into the high-speed traffic where several were struck and killed. The crows then harvested the roadkill. David Perkel, a professor of neurobiology at the University of Washington, told us about

In Tasmania, forest ravens eat more than roadkill. These innovative foragers quickly have learned to check the pouches of killed marsupials just in case a baby is inside. Here a forest raven robs the pouch of a Tasmanian devil.

a crow he watched in close pursuit of a gray squirrel. The squirrel dashed across the road and was struck by an oncoming car. The crow waited for traffic to clear and then dined on fresh squirrel. Perhaps this crow, and others behaving similarly, just caught a lucky break, but lucky breaks are remembered and perfected by sentient animals. If luck becomes predictable, crows may run squirrels more frequently, especially during rush hour. Certainly the connection between the chase and the kill is easy for these birds to comprehend.

Causal connections lead corvids to spend time window-shopping to get what they need. On Pender Island, just off the west coast of Canada, ravens chase robins into Noni Pope's windows. The stunned or dead thrushes are quickly retrieved and eaten. Crows use the same bull-rushing tactic to stampede sparrows in Seattle; sometimes they even miscalculate and hit the glass full steam. A pair of ravens who live just east of Hitchcock in the North Cascade Mountains used teamwork to flush red crossbills into the windows of buildings. The crossbills become superabundant in this area when conifer-seed crops farther north fail, as happened in 2008 around the complex of buildings known as the North Cascades Institute (NCI). Elvis, a male raven whom John trapped and tagged the previous year on a large slab of bacon, would rush the crossbills down a corridor between buildings, and his mate would intercept their progress and turn them into the glass windows. Stunned crossbills littered the ground afterward until Elvis and his mate grabbed mouthfuls to eat and cache for later dinners. Biologists at NCI estimated several hundred crossbills met their fate that year at the brain and beak of the ravens.

Elsewhere, crows use bait to catch a meal. In Tokyo, at least one jungle crow distracts dogs in order to steal their food. Cindi Sonntag watched a crow perch on the edge of her dog's kennel and drop something over the edge as her pet was dining. As the inquisitive dog ran to mouth the item, just beyond the fenced perimeter, the crow swooped in and got three mouthfuls of kibble from the dog's bowl. The use of lures is perfected by pet corvids. In eastern Washington a wily crow waits near an open window and drops bread on the sill. This bird is not looking to fly the coop; it is waiting for a sparrow to come in for a snack. When the sparrows get hungry enough to come within striking distance, the crow instantly snatches one. A twelve-year-old raven in Alaska with a meal on its mind also lures others to their deaths. This hunter puts food just outside its aviary wall and specializes in catching Steller's jays that flock to a seemingly free lunch. Sparrows and juncos, species with much smaller brains than jays, apparently are not challenging enough for this wise old hunter. Tony's raven, Macaw, routinely flew across to the neighbor's to harass his dog at feeding time and learned before long to distract the canine with a pull on the dog's tail. The dog would abandon the food bowl in favor of a chase and, once the dog was out of the backyard, Macaw would quickly fly back to gorge from the dish as the hapless dog was still wandering the street.

The ability of crows and ravens to adjust their behavior after considering what they sense and what they expect—the curious mulling

Macaw pulling the tail of a husky.

over of a situation that is enabled by the neural circuits in their large brains—underlies the innovative feeding tactics that are so common in this group of birds. Working with accomplices opens up even more possibilities. As a rule, corvids mate for life. Constant, close contact for years, even decades, enables these pairs to coordinate and reciprocate to expand their palates in ways that are impossible for a lone bird. Elvis and his mate were extremely effective at hunting crossbills because of teamwork. In Juneau, Alaska, one pair of ravens worked together to get whipped cream out of a high-pressure can. One bird gripped the head of the can with its feet and used its beak to push the button atop the can and squirt out the cream. As this bird dispensed the treat, the other ate. Cooperatively, they switched positions. In eastern British Columbia, raven pairs course over grasslands in search of small mammals; as one bird flushes a rodent, the other dives down for the kill. Closer to the coast, pairs of ravens cooperate to herd Jeff Williams's chickens into a fenced corner, where they quickly dispatch and eat the meaty birds. Even young seals are no match for a pair of ravens. Early Arctic explorers observed a pair of ravens circle low over a seal pup sunbathing on an ice pack. As the birds drew close, one landed over the seal's breathing hole, blocking its escape route. With no means of escape, the pup was an easy target for the other raven, which attacked with skull-crushing blows. Both shared the meal.

Ravens use teamwork to block the escape route of a seal pup.

Small gangs of corvids seem unable to resist pulling the tail of a more powerful animal. A raven will frequently pull an eagle's tail when both gather around a spent salmon or frozen moose. A trio of arctic ravens stole a meaty bone from a sled dog. As two birds cawed hoarsely at the dog's head, a third flew down and yanked hard on the canine's tail so that the dog dropped the bone. The confused dog did not even see the other ravens fly off with their prize. On a humid summer day in Wisconsin, Tom Gehring heard a raven commotion and caught a group of four birds, likely a family, pulling the tail of an adult porcupine. As the spiny beast charged one bird, another rushed from behind to pull its tail, causing the confused rodent to spin around and confront the new offender. Back and forth the defensive animal charged; for fifteen minutes it could not escape. When the birds spied Tom and flew off, the adult porcupine was spared, but a nearby, nearly dead young porcupine seemed in shock. It had severe trauma to its tail and back; certain prey of the small unkindness of ravens. In Florida, crows also pinch the tails of formidable mammals, river otters, to steal a meal. Sometimes they may do it just because they can.

This propensity for tail pulling seems to be a worldwide inclination among the larger corvids. The great white-necked ravens of Africa are described by Friedrich Von Kirchhoff as routinely stalking his pet meerkat to approach unseen and tweak its tail. Given the scale of this raven's beak, one imagines it was more of a crunch than a tweak.

A white-necked raven in Africa pulls a meerkat's tail.

While impressive, the joint ventures of corvids do not seem to be on par with the cooperation practiced by people or our closest primate relatives, the chimpanzees. To date, however, only one study has explored their cooperative behavior, using for its subject the rook, an extremely social corvid, about the size of a large crow, which eats, sleeps, and nests in rookeries of several hundred individuals. Flocks of these all-black birds with rather bare faces pluck insects from fertile farm soils throughout a large part of northern Europe. In an experimental laboratory in Cambridge, pairs of rooks, especially those who had already established good relations, quickly learned to cooperate with each other to pull a food tray into their cage. To succeed, each bird had to grasp a short string and coordinate its heave and ho to slide the tray and its bounty into the cage. They were as successful as chimps at this task, although the birds seemed to succeed with less than a full understanding of the task. If one rook was allowed in the cage ahead of its partner, it did not wait for its assistant before impatiently working at the string. A lone rook with experience teaming up for success seemed blind to its need. Chimps, on the other hand, did not fiddle with strings that required teamwork until their partner was able to do likewise.

The engine of innovation is the willingness to try something new. Sometimes this can lead crows to smoke and drink. Margaret Winnie's pet crow landed on her dad's shoulders to steal cigarettes. Indian house crows in the Maldive Islands also steal cigarettes. Thus far there are no reports of crows lighting up, but even ingesting nicotine could be problematic. Chickens fed nicotine forgot much of what they had learned and often collapsed into a sleeplike stupor.

Crows may prefer drink to smoke. Jay Bennett lived with a drinking crow in the 1970s near Tacoma, Washington. Jay's pet made a habit of eating fermenting cherries from the backyard. The more the crow ate, the drunker it got. The drunk bird staggered, called wildly, and flew. Jay's son Chris vividly remembers the erratic, swooping, twisting flight of the drunken crow. Wild birds also drink. On Pender Island, British Columbia, ravens steal full beer cans, pierce them open, and guzzle the brew. Alaskan ravens may drink to stay warm and alert on cold mornings. They are quick to grab a discarded coffee cup, especially those bearing the Raven's Brew logo. The powerful bird carefully, even gently, grips the rim of the cup and tilts it back so the stimulating brew drains efficiently into its expectant maw.

House crows steal cigarettes.

A raven in Alaska sips the dregs of a cup of *Raven's Brew* coffee.

To us, it was clear that Hitchcock "the wiper swiper" was motivated by a combination of circumstances and experiences—territory defense, frustration, food search, and innovation topped our list. This superbly adapted bird was simply exercising its talents in the wrong place. There may have been some other prompting for Hitchcock's madness, but we saw nothing to suggest Hitchcock was actually using the stolen rubber—no nest reinforced with pliable wipers or lined with rubber sheeting. We dreamed of uncovering the first raven ceremonial display or tool use that featured wipers but saw nothing of the sort. Hitchcock left most of the shredded rubber on the road near each vandalized auto. To Hitchcock and the other rubber-thieving birds we've heard about, the reward is in the chase, not the prize.

Tony's experience with his adopted raven, Macaw, suggests that sometimes ravens simply enjoy mischief. Each morning the bird was free to fly about the neighborhood of Edmonds, Washington, to satisfy his needs for exercise and, we suppose, his insatiable need to make sense of his world. He rarely ventured beyond the range of Tony's voice or shrill whistle. Whenever Tony would produce his best high-pitched shriek, the big bird's black form would burst shortly from the distant trees and set sail down the middle of the road in his direction, anticipating a breakfast reward. One day, after Macaw had been out for an hour or so, Tony heard a commotion at the end of their block suggesting distress. Fearing the worst, Tony whistled, and the formidable raven quickly responded as usual, flying directly to his front lawn. Landing in front of Tony, he ambled over in a lordly walk expecting to be fed, when Tony noticed a clothespin clasped firmly in his beak. A "trophy" so to speak, from his morning's adventure. Apparently he had been wrestling with it on a clothesline when Tony called him home. Tony took the pin, which Macaw released willingly, and provided him with a handful of a favorite meal—chicken backs and egg yolk.

On the following morning, Tony followed his usual routine of opening Macaw's enclosure door to allow Macaw to hop about and begin exploring. On this morning, however, Macaw had a different agenda as he bolted out to fly directly to the spot he had emerged from the day before. Within seconds Tony could hear his "croaks" and "krawks" and assumed, with some trepidation, that he was back at the clothesline in which he had taken interest the day before. Tony immediately started whistling to get him to return, but Macaw was faster in applying his mischief. In he flew with an even bigger prize dangling from his beak and in stark contrast to his ebony plumage—a pair of ladies' underwear.

Macaw with pilfered underwear.

Pet ravens from Michigan and Arizona also embarrass their owners by raiding the neighbor's clotheslines. Some wild birds take this behavior to the level of a rampage. Darla Dehlin lives in northeast Washington, near the Canadian border. A large flock of ravens terrorized her ranch: killing newborn goats; stealing the dog's food; removing chinking from her cabin; stripping rubber from her camper; pilfering tools, mail, and a cell phone; destroying cans of soda, oil, and fuel; dismantling a snowmobile; and taking any clothing hung out to dry. Anyone missing socks?

Crows enjoy a good trick. Daryl Clark was talking to a friend just outside his home when a tossed apple barely missed them. About a week later, again Daryl was nearly beaned by an apple flung from his roof. In both cases, Daryl looked up to see a crow looking down and then backing slowly out of sight. The bird was quiet, perhaps wondering how to improve its accuracy.

Sometimes delinquent birds seem unable to get enough mischief. Golfers near Leavenworth, Washington, were upset when a crow stole a bagged sandwich from their cart. When a crow returned with the now-empty bag and replaced it in the cart two holes later, they were dumbfounded. Similarly, in Barclay Sound kayakers were upset when ravens stole a fresh pie, and were really angered when the pair returned with the pan the next day and dropped it on the boatmen. Could these birds have actually made the connection between the container and the meal? Were they asking for second helpings? The directed nature of the behavior—returning the empty bag and pan to appropriate people—suggests the birds had made such a connection. Birds routinely return to places where they successfully found or stored food, so relocating and following a reliable source of food is absolutely normal. Corvids have a long association with people, and it is most certainly based in large part on our being seen as a source of sustenance.

Occasionally a crow will do the right thing. Delbert Wichelman grew up in pre–World War II rural Minnesota. As was common at the time, he raised a pet crow, Jimmy Crow, who regularly stole laundry and nearly anything else he could beak. It was fun until Delbert's sister tempted Jimmy with her new diamond engagement ring, which he snapped up and, flying to the neighbor's roof, cached it in the high rain gutter. The kids, worried where the crow might move the prize, called at Jimmy to return it—please. And he did, grabbing it and flying back to the sister, dropping the ring at her feet.

A similar good deed befell Suzanne Wyman as she lunched on a dock high above Cowichan Bay in British Columbia. Suzanne wears dentures and had taken them out to enjoy every good taste of her lunch. As she brushed the crumbs off her newspaper, she accidentally swept her dentures off the dock and down to the mud flat, thirty feet below. Distraught, but unable to retrieve the teeth, Suzanne returned home and eventually bought new chompers. A few weeks later, her neighbor called to ask if she might have lost her false teeth. The crows it seemed had left an unusual gift on the neighbor's dock. Indeed, they were the teeth Suzanne lost weeks earlier, but rather than muddy and worn, they were polished like new. It's reasonable to consider the motivation of the pet crow Jimmy as following commands and pleasing the master just as would a dog. But the motivation of the Cowichan crows is difficult to understand. Perhaps they simply tired of a unique find and by chance left the dentures on a nearby dock.

> From Twin Falls, Idaho: CITY VOTES TO SHOOT, POISON TROUBLESOME CROWS
> From Los Angeles: FLOCK OF TROUBLE
> From Berlin: CROW-ATTACK SEASON HAS BERLIN ON EDGE

The innovative mischief that is so deeply a part of the mystique and wonder that endear crows and ravens to many is upsetting to others. The headlines above are but a tiny sampling of the ire crows and ravens often engender. Concerned citizens, cities, even nations plan and carry out wholesale corvid slaughters. In the United States alone, for example, the federal government killed more than 10,000 crows and 3,800 ravens in 2009. Living with these potential winged troublemakers is challenging, frightful, and even costly to many people. But we believe that people should try to discover and address the root cause of corvid trouble before they pull the trigger. This is ethically appropriate because the crows and ravens that challenge us are sentient beings; we should at least return the favor of using our thought processes.

The populations of crows and ravens, and the trouble they bring us, are increasing in many parts of the world. This is a predictable response to human lifestyles that directly supplement these birds with food, water, and shelter. The first step in lessening problems is simple: understand how we inadvertently supplement corvids and actively reduce those subsidies. In the United States, three cultural shifts are restoring a more natural balance between corvids and their foods, and therefore between corvids and ourselves. First, many open garbage pits, which concentrate and increase corvid populations, are being closed. Second, the hazard of highways to many animals is increasingly recognized and in some places reduced with designs that accommodate wildlife movement. This is important for human safety, but it also reduces a significant food source (roadkill) for corvids. Finally, wolves are being restored to our wilder ecosystems. These carnivores attract corvids who scavenge on the remains of their kills rather than on the supplements of our rural lifestyle.

Lowering food subsidies and restoring natural food webs is a critical first step in reducing the problems that corvids pose to society. Reducing unnatural sources of water in dry environments and rethinking man-made structures that encourage nesting in terrain devoid of natural corvid nest sites (cliffs and trees) is also needed. Some local efforts to accomplish this are underway. Part of the process for limiting these clever, exploitive species is to take a bit of time to pick up, put away, and recycle our refuse from camps, picnics, and drive-in eateries that host a legion of foraging crows. Install crow-proof bird feeders and make them friendly only to the species that require them. And of course keep the lid on the garbage can—a plastic bag is an open invitation for these ebony bandits to surgically slash, sample, seize, and dash away with everything within.

After subsidies are reduced, other changes in our lifestyles will significantly lower the trouble that corvids can bring. Protecting young farm animals in covered sheds and encouraging shrubby, unkempt places in our yards where potential prey can escape, would reduce the ability of crows and ravens to prey on other species we value.

Even raising peacocks can make a difference. Peacocks were part of the solution that Darla Dehlin used to disperse the flock of ravens that was dismantling her ranch. The peacocks cleaned up spilled grains and pet foods that were encouraging ravens to loaf on Darla's property. No subsidies, no ravens. Limiting our use of natural resources before demanding that other species do so would also be prudent. Yet in many parts of the world, we kill crows and ravens simply to increase our share of the ducks, grouse, and quail that corvids naturally prey upon.

Lifestyle changes will reduce the problems we have with corvids, but their innovative ways will require some innovations of our own. Aversive conditioning, as we used to reform Hitchcock, or fencing off the nests of sensitive species, as is done to lower crow predation on threatened piping plovers, may provide some help. Even tapping in to the natural tendencies of corvids to steer clear of foods that make them sick may be effective. In the redwood forests of Northern California, researchers are teaching Steller's jays and ravens to stay clear of the eggs of the endangered marbled murrelet, a small seabird that lays a single egg on the broad, outstretched, mossy branches of old conifer trees that used to soften the rugged Northern Pacific coast. Once plentiful, the forests that provided these unusual nest sites were leveled by an expanding American human population. Fewer nests and more corvids are threatening the seabird's existence.

Park managers, interested in conserving murrelets, figured corvids could learn to stay away from the briny seabird eggs simply by flooding the market with look-alike eggs packed with a raunchy emetic. As we write, the forests are full of tricky eggs, teaching nest predators to leave the distinctive large and speckled murrelet eggs alone. The reasoning is sound; educated to the possibility of a distasteful meal, long-lived corvids will resist murrelet eggs, and their fierce defense of territory will keep naive, would-be nest robbers away. Better to teach than to kill, and have to keep killing.

The mischief of corvids is fascinating and engaging, but some view it as an intolerable delinquency. For the sake of control, this behavior should be better understood. Seeing that delinquency is rooted in the mental abilities of remarkable animals that survive because they can learn quickly and must fiercely defend their territory adds a perspective that calls us to look inward. From past experiences we know we cannot afford to routinely apply the frontier strategy, that is, killing an irritating or competitive member of our ecosystem—there are too many unforeseen consequences. Better choices require some thought and sacrifice but are respectful of other thoughtful and innovative animals with which we live. And ultimately our tough choices will sustain our natural diversity. In areas where we dominate the ecosystem—our cities, suburbs, and farmlands—many native crows and ravens can thrive. In a world where so much of our natural heritage is being lost, why not celebrate the few bright spots where it is surviving and adapting? A headline from Cape Cod suggests some of us do: CROW POISON PLAN KILLED.

5

Insight

Al, a bold young raven in British Columbia,
takes a turn at pulling up cheese.

THE CENTRAL Pacific coast of British Columbia is one of North America's great remaining wildernesses. Near-daily rains soak the rugged terrain, feeding unsullied rivers on short runs to deep fjords that snake among rafts of uninhabited islands. Cold rainforests of massive trees—hundred-meter-tall Douglas-firs, western hemlocks, and western red cedars—cover the terrain with an impenetrable, verdant cloak. Massive tree trunks burst from thickets of leathery salal, prickly salmonberry, vengeful devil's club, and bountiful huckleberry bushes. There are no roads, and the few walkable paths are passages that have been steamrollered and maintained by bears, both black and grizzly. Abundant wolves join eagles and ravens to feast with the bears on wild salmon returning to spawn. This wilderness ecosystem is intact, though frayed around its edges.

Our good friend Tom Rivest and his wife, Marg Leehane, run a small floating retreat, Great Bear Lodge, anchored deep in the British Columbian wilderness. Ecotourists visit Tom and Marg to learn about the rainforest and its bruins. John finally got a chance to join them in July 2010. A friend of Tom's from graduate school and a long-term research partner of John's, Marco Restani, organized our expedition to the Great Bear Rainforest to experience its solitude and observe its wild residents. We headed north from Seattle with biologist Nick Mooney from Tasmania and Professor Johann Kopel from Berlin, crossed the Salish Sea aboard a ferry, and drove all day to the northern tip of Vancouver Island. From there, in the village of Port Hardy, we hopped a small floatplane for a forty-minute flight to the Great Bear Lodge. A high ceiling of clouds allowed a direct flight as we buzzed low over graveled beaches, drifts of massive logs, and the near-vertical slopes of old-growth tree stands scarred here and there by clear-cuts that testified to the consumptive demands of society. We followed Smith Inlet east, thirty miles deeper into the throat of this continental wilderness, to where a river poured into the still sea and bright green paddies of wild Lyngby's sedge rimmed an estuary. We dropped, circled, and skimmed to a landing on a runway of water.

Within an hour of settling into the lodge, a raven adopted us.

Thin but resolutely brave, the recently fledged bird walked up the gangway from shore to the lodge like a guest searching for the concierge. The local raven pair was tending its annual family of three loudly begging fledglings. The bird that joined us may have been the weakling of that brood or a recent immigrant from a neighboring clan. The resident adults berated the young raven with threatening calls and exhausting chases. Apparently the lodge was the nearest safe haven. The timing for us was fortuitous, if not mystical. What better visitor to join a lodge full of biologists than a raven? We named the bird Al, and within twenty-four hours Al was standing on our laps recycling pancakes, grapes, cheese, salmon skin, and apples. In the process, he laid bare how ravens so easily inserted themselves into the indigenous human culture of the Pacific Northwest. A shaman only had to offer a young raven a bite to eat, and almost immediately human and bird forged a bond that would only strengthen with time.

We weren't looking for a feathered oracle and in fact felt guilty for feeding this wild beast and perhaps making it dependent on people and endangering it in the future. But the implications were stunning. Al was only a month or two out of the nest, likely off his natal ground for the first time, and, instinctively, he sought shelter with people. Indeed, our species have mutually shaped each other. Being curious scientists, we leveraged this fortunate encounter and devised simple experiments to test Al's mental abilities.

A decade and a half earlier, Bernd Heinrich, our colleague from the University of Vermont, had given five hand-raised ravens a battery of tests examining their problem-solving abilities. To probe the ravens' ability to use insight, Heinrich hung chunks of meat from strings off the ravens' perches. To obtain the meat, the birds had to precisely link together a series of difficult actions. The ravens might have already known each individual action—reaching down and gripping the string, pulling up a loop of string, setting the loop on the perch and stepping on it, releasing the string from the beak, reaching farther down the string for a second pull—but orchestrating the individual moves into a profitable sequence that obtained the prize was entirely new. They could have solved this new problem by assembling old actions into a new routine through trial and error or by understanding the full task and using this insight.

Heinrich's ravens varied tremendously in their problem-solving abilities. Two birds mastered the task in only a few trials. Two others consistently pulled up meat within a week of their first attempt. One bird never succeeded. This variation rules out the idea that string pulling is instinctual, because all ravens don't have the instinct. Trial-

and-error learning also was not obvious: while the time to master the task varied, this delay wasn't because of clumsy practice. Once a raven stepped on a pulled loop of string, however, he flawlessly exercised the full sequence of pulling, stepping, releasing, and reaching. Heinrich's birds may have learned with experience to speed up the process, but they seemed to understand the required task and to succeed at it by instantly piecing together a long sequence of actions to reach the desired goal. Further evidence that the birds understood the problem at hand lay in their unwillingness to pull up rocks or overly heavy foods like sheep heads when those were on the strings. Nor did they try to fly off with their leashed bits of meat; when frightened, they dropped the meat, then flew. When Heinrich crossed strings so that the raven pulled a line situated above the meat but retrieved a rock, one bird quickly solved the task by eventually selecting only the string attached to the meal, but three others never did—nor did they learn from their errors.

Back at the Great Bear Lodge, we tied a chunk of cheese to a twenty-inch-long piece of string hung from a small patio roof. Al flew right up to the roof and walked toward the string, full of confidence. He looked over the edge at the dangling cheese and immediately reached down and beaked the string, pulled it laterally along the roof, stepped on the slack, and released his bite. In seconds he repeated the sequence three or four times and claimed the cheese. Brilliant! Surely this wild, young raven deep in the Canadian wilderness had never seen string, let alone a dangling bit of cheese. We had heard of ravens in Alaska untying fishermen's stringers to steal their catch, but there were no fishing people up here.

We quickly rebaited the string, and Al immediately retrieved a cluster of grapes. Then another one. Without hesitation and with near perfection, he seemed to understand the problem and employed a neat repertoire of moves to get a meal. No trial and error here. Al seemed insightful, especially when we provided a large hunk of meat on an extra long string. Rather than haul up this prize, Al grabbed it from below and pulled it from the string. But when we crossed strings, Al was clearly confused. As he pulled one string, he stepped on the other and got neither meat nor rock to the roof.

Al's motivation to explore, to innovate, to assess, and quickly solve a novel problem proved to us how gifted ravens are. Impressive mental skills allow ravens to succeed where most other birds fail. But a simpler cognitive mechanism—one that does not require insight and planning—might explain Al's ability: the rewarding nature of each step of the process. A world away, New Caledonian crows were pulling strings of their own to provide the answer. This species of habitual tool users, the tribe whence came Betty the wire hook maker, quickly mastered the string-pulling test and directed their behavior to efficiently attain a goal. Three of four crows used the reach, pull, loop, step, release, reach sequence to garner meat on their first attempt. The fourth bird got it on its second attempt, and the total time for all four birds to solve their first-ever string-pulling exam was an astonishing sixteen seconds.

Impressed, the New Zealand researchers who conducted these studies wondered if the reward of seeing the food moving toward the crow might be enough in itself to motivate the next sequence of behaviors that leads to complex string pulling. Rapid learning of an association—pull equals closer food—would depend on the cognitive capacities corvids possess. The strengthening of synapses that connect neurons carrying sensory information from their mouths and feet with neurons transmitting visual information might occur in the forebrain's executive or integrative center or in the hippocampus. Neural loops between the forebrain and thalamus would allow the mental expectation of how the lassoed food responds to the bird's actions to be compared to the actual way the string pulling affects the trajectory of the food. Immediate and ongoing mental assessment could allow rapid, on-the-fly sculpting and fine-tuning of the birds' motions into a flawlessly executed behavioral sequence. This behavior is extraordinarily cognitive, to be sure, but it may not be reliant on full, immediate comprehension of a task and planning; the successful string puller may only need to make connections between its actions and an ongoing rush of good news—the food is approaching as expected by my action—to solve the problem.

To know what is in the mind of a string-pulling crow, our colleagues in New Zealand used an ingenious apparatus to influence the immediate feedback to string pullers and therefore comprehend the degree to which insight versus observing and adjusting are used to guide pulling. The group of four experienced string pullers who solved their first exam in sixteen seconds was given a new test where the string holding food was suspended through a small hole in the floor of their cage. If these birds used insight to understand the task, then they should reel up the string, even though seeing the food at the end of the string was difficult. They did not. The error rate of these competent pullers increased tenfold, and they secured the meat on only half of their attempts.

The few successes scored by this group of crows seemed to rely on careful peering through the hole at the food during the retrieval period (they could have glimpsed the food through the hole prior to the start of the experiment to know a meal awaited). Even this poor performance was better than the near complete failure of totally naive string pullers on the test; only one of four crows whose first experience with suspended meat was with it hanging through the hole in the test chamber was successful, and this bird succeeded only in half of her ten trials. But another group of naive crows was given a visual aid—a mirror in which they could watch the reflected food rise up toward them in response to their string pulling. The mirror itself freaked out two of these birds, but the other two faced the mirror to successfully solve the problem, each doing so on six of their ten trials. Overall this group of crows was as successful as experienced birds and twice as successful as other naive birds at pulling up the food through the hole. The immediate sensory feedback—action matching expectation—seemed to enable the complex string-pulling behavior.

Despite a small sample of test crows, this imaginative experiment shows that behaviors that seem to depend on insightful understanding of a problem may in fact be guided by more direct sensory feedback. The tendency of corvids to test their environment with expectation and then to quickly associate success, even partial success, with their explorations may be a key learning mechanism. Corvids seem to possess goals and direct their actions toward them, but they need not work out in their minds fully how to obtain the goal. They only need to try something—often a new innovation—and then shape their behavior to the immediate result of their action. This makes absolute sense to us; we see our familiar crows, ravens, and jays always testing and probing their environments and then adjusting their behaviors to fine-tune their responses and better exploit the resources of their dynamic environments. We humans also regularly act without extensive planning and learn from the consequences of our actions.

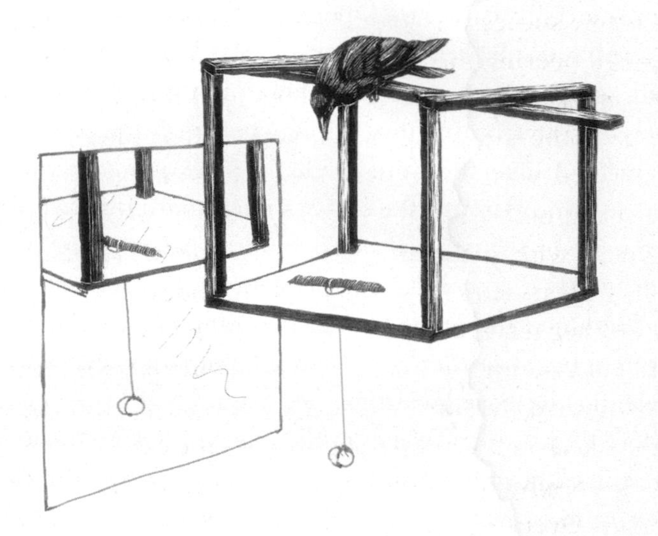

Testing insight versus continuous reinforcement in the string-pulling behavior of New Caledonian crows. Pulling efficiency was greatly aided by a mirror in which the birds could see the approach of the reward, suggesting that insight was less important than rewarding feedback to this behavior.

The crows that hunt the exotic grounds of Seattle's Woodland Park Zoo track the variable bounty served up daily for the captive animals. Having already made associations between the curious actions of visitors, pilfering their fries and sandwiches, these crows are now cuing in on the zookeepers. The Humboldt penguins thrive in their new, natural exhibit, which features rushing water and a weekly feeding of live trout. Each Friday, just before noon, the zookeepers heft buckets of fish over to the giant aquarium and pour them into the water. For a moment the young trout cruise in apparent freedom, and then the hungry penguins hit the water to torpedo in pursuit and deftly snatch one tasty morsel after another.

It is predator versus prey here as it is in the penguins' native Antarctic waters. Except that here, another clever predator also lurks. Anticipating the weekly feeding ritual, a dozen wild crows have assembled among the faux cliffs ringing the penguin pool, expecting a trout delivery. Seattle's penguins have been eating live fish for only a few months, but already the crows know the routine and beat the keepers to their posts by a few minutes. Synapses among the neurons in their brains link timing cues with the reward of a nutritious meal. And as the trout evade the sleek penguins and seek calm waters in which to rest, the crows snatch them from above. Like herons, the crows thrust their beaks into the water, grip a wiggling fish, and promptly dispatch it with a violent thump against a rock. Crows have fashioned foraging strategies to match the varied habitats the zoo offers because they associate reliable cues, even precise times, for feedings in these locations.

More than two and a half millennia ago, Aesop, the ancient Greek slave, knew about the innovative problem-solving abilities of crows. One of Aesop's fables, "The Crow and the Pitcher," tells of a thirsty crow that dropped stones into a vessel to raise the level of water up to the point where the bird could dip its beak for a drink. Obviously, the crow knew the goal—raise the water—but did it carry out a detailed plan gained by insight, or did it try something and then build on the results of an innovative stone drop? As the string-pulling ravens were rewarded by each movement of the food closer, did the thirsty crow notice the water rise higher after the stone drop?

Those rooks from the previous chapter, who did not quite grasp the details of cooperative string pulling, provide an answer. The rook pokes its long beak deep into the earth to seek and grab each meal. It is not known to use tools as it probes tilled soils for worms. However, a few captive rooks have become tool-wielding disciples of Aesop.

The birds use stones to raise water levels to a point where succulent waxworms float up to within reach. Immediately or by their second experience at the latest, without prior training or practice, these rooks dropped just the right number of stones—one by one—into a tube of water to obtain a meal. Regardless of how full the tube was at the start of the trial, they made no attempts to grab the worms until the water was sufficiently raised. This suggests a capacity for insight as the birds quickly switched from small to large stones to accomplish the task more quickly.

Clearly these rooks understand the goal—get food—and the problem—displace the water—and solve it through a series of unique behaviors. The full and precise sequence is remarkable and illustrates that the birds can work toward a distant goal. But immediate feedback, not insightful planning, seems to be the cognitive process guiding rooks. They simply adjust their actions to each successive, stone-resultant rise of the floating worm. In a control test where the tube had sawdust, not water, the rooks also dropped stones. Like the crows that stopped pulling strings when they could not see a tasty treat getting closer, these rooks quickly stopped dropping rocks in the sawdust that was not displaced by their efforts. Either the birds understood that their plan was foiled or they lacked the immediate reward necessary to motivate a complex behavioral sequence.

Rooks demonstrate Aesop's fable by adding pebbles to a column of water to float food to within their grasp.

The innovative rooks tested all aspects of their environment in highly unusual ways. They developed tool use to solve novel problems. They learned physical relationships by responding to the actions their tools produced. By attending to the results of their innovations, they worked just long enough and used the best tools for the job to obtain a meal. In other experiments they even mastered the crafting of hooks from wire, just as Betty did, to retrieve out-of-reach buckets of food.

When it comes to scoring a meal, it seems corvids will try just about anything. And by careful observation and consideration, they quickly transform what is initially an exploratory guess, albeit a guess educated by past experience and emotion, into a novel solution to a new problem. The goal of obtaining food seems foremost on a crow's mind, and any action stands ready to be shaped, by a forebrain quick to associate action with reward, into an efficient and elegant solution to a complex problem.

Does this basic learning strategy—rapidly associating action with reward—really explain all that we see corvids accomplish? What seems insightful, the gathering and stacking of crackers and chips for efficient transport rather than ferrying one at a time, dunking dry foods in birdbaths and gutters to soften and hydrate them, even carefully carving off corners of suet so that large chucks, not small flakes, can be gathered, are easily explained by rapid association with immediate, if partial, reward. Insight and planning cannot be ruled out, but they are not needed to explain these amazing acts.

In other situations, however, insight seems absolutely required. The latest test of the string-pulling New Caledonian crows may finally prove the use of insight by a bird. Pulling a string or dropping a stone for food is immediately reinforcing and, as we have seen, may be quickly perfected for that reason alone. But pulling a string to get a key to unlock a box to get a tool to get food is not so simply reinforcing. This requires the suspension of an immediate return on the bird's efforts. Especially if pulling the string for the key in the first place is known to be a waste of time. New Caledonian crows seem to understand such complexity, then plan and act appropriately to score a meal.

Researchers challenged crows in a three-stage problem that required the birds to get food by first pulling up a string that held a short stick tool, then using the short tool to get a longer stick tool, and finally using the longer stick to push food out of a hole in a box.

Captive crows were trained in two ways. One group was trained to do the many individual behaviors that would later become components in a complex sequence: pull strings for food, pull strings for tools, use short tools to get long tools, use long tools to push food out of a box. Another group was trained only to pull strings for food and to use stick tools to access food. This group also learned that short tools were ineffective at obtaining food from the box.

Both groups of crows passed the three-stage test with little effort. Each of the three crows that knew the many components solved the task the first time they were examined. On average each bird looked at the crazy array of sticks, strings, and boxes for thirty seconds, then flew right to the perch, pulled the string to retrieve the short tool, took it to the box to get the long stick, and then used the long stick to push out the food. All four of the crows that knew only generally about strings and sticks also solved the problem; two did so on their first attempt.

What is so important about these four geniuses is that they had previously experienced consistent failure with the short stick; it never enabled them to get food. They had stopped bothering to retrieve it prior to this exam. But now, in a new setting, they immediately recognized that this formerly ineffective tool would be useful in obtaining the appropriate long tool. Pulling up a previously ineffective tool could have been self-reinforcing, but *only* if the crows already had a plan to use this tool to get another tool to get food. Pulling up the short tool did not directly move the food, and in all past experience with the short tool it never even enabled the acquisition of food.

These birds organized a sequence of behaviors to follow a mental plan. They demonstrated insight—the ability to see far enough into a situation to fully grasp the problem and the solution. Corvids can use at least two complex cognitive processes to reach goals that they understand. Rapid associative learning shapes simple actions into complex behavioral sequences, and insight allows plans to be crafted from memory, appraised mentally, and carried out.

The combination of association and insight may explain the uncanny abilities of crows to anticipate future events. Our lives in the Pacific Northwest often include lengthy waits to board one of the

True insight by corvids was demonstrated by New Caledonian crows' responses to a three-part intelligence test. The crows had to retrieve the short tool to obtain the long tool that could access the food.

many ferries that usher cars and people across Puget Sound. While so engaged one day, Tony observed a curious feeding strategy. The crows were flying up from the beach below where, at low tide, they were picking up small clams and then dropping them on the empty roadway adjacent to where we were parked. Their drops were not random as this was the now-open lane upon which the cars from the arriving ferry would disembark. No cars currently traveled on this road to cue the birds' actions. One after another a closed shell was dropped until there must have been several dozen scattered on the road's surface. While some of the shells were surely cracked, none of the birds flew down to investigate but returned to the beach to retrieve more clams. After thirty minutes or so the incoming ferry arrived, and a hundred cars rolled from the boat, crushing the clams as they did so. When the ferry was empty, our lines of waiting vehicles began loading from the opposite side of the road, and Tony watched the crows descend to the traffic-free asphalt now littered with clam meat.

Associating an easy meal with a specific time is learned by consistent experience. While appearing to involve active planning for a future event, anticipating a reliable event like a zoo feeding or the traffic from a ferry can be accomplished through relatively simple associative learning. Though exercising considerable restraint, crows don't have to think about past rewards and imagine them happening in the future to learn that piling clams on the roadway after a ferry leaves will be rewarded in an hour. They may have such forethought, but it is simpler to assume the synapses between neurons that link reactions to the sight of exposed clams at low tide, the locations and timing of a particular road free of traffic, and the picking and eating of clams have been strengthened by the reward of a bountiful meal.

The ability to think about the past and plan accordingly for the future, projecting oneself forward in time, is considered by many to be a unique feature of humans. The impressive, integrative prefrontal cortex of our brains may enable this handy trick. But given our new understanding of birds' brains and the apparent way in which their executive center (for instance, the NCL as illustrated in chapter 1) works like our prefrontal cortex, it seems at least possible that birds like corvids may employ true mental time travel. Our smugness at being the only mental time travelers may have a lot to do with the fact that we can tell each other about such experiences, but we cannot talk about this possibility with other species. An ingenious set of recent experiments with western scrub-jays indeed strongly suggests that corvids are actively prospective, future planners.

Scrub-jays are small corvids that, like all members of this brainy group, cache food when surpluses exist. Caching itself seems to imply planning for the future, but it may be motivated and enabled simply by a current need—food—and well-developed spatial memory. To rule out the role of current need, a research team at Cambridge University tested jays on their abilities to plan for breakfast. In captivity, eight jays quickly learned (in only three trials) that one compartment of their cage would have breakfast, while another compartment never would have morning food. When given an evening meal of thirty whole pine seeds and free range to visit both compartments in their cage, the birds did an interesting thing. They ate a few seeds and then set about caching the others. After thirty minutes they had cached three times as many seeds in the compartment in which they had previously learned no breakfast would be served than they cached in the compartment where breakfast was always provided. It did not matter what type of food was presented or previously consumed in a particular compartment. Additional tests with dog kibble and peanuts confirmed that the jays associated future hunger (not the past consumption of a particular food) with the need to cache in a specific place, at a particular time. Scrub-jays anticipate a future need—a morning meal—with a particular place, then develop and carry out a plan to remedy that need.

Clearly, this is another example of truly insightful behavior. One more trait we held as uniquely human—the ability to anticipate and prepare for the future—is shared with our feathered companions. Maybe those crows loitering around Puget Sound ferry docks, stockpiling clams, really do imagine themselves enjoying a future feast.

Crows and ravens make many decisions demanded by innovative lives that seem to include more than simple associations and insights into physical relationships. Crows also appear to know what other animals in their world are thinking, and they use this information to inform their behavior. In Florida, for example, crows cache clots of blue-gray clay in nearby meadows. There is no immediate reason these birds should do so, but there are intriguing possibilities. Lawrence Kilham observed that the balls of clay were about the same size as the heads of catfish he'd often seen crows scavenge. When these crows feed on catfish heads they are harassed by a resident red-shouldered hawk intent on stealing the prized morsels. Kilham thought the clay might be a decoy acting to deceive the hawk into ignoring the crows.

In the 1950s Jimmy, a pet crow from Oregon, also had a penchant for caching seafood. Jimmy lived with the owner of a seafood market and hid all manner of fishy items in the lawn behind the shop. The keeper's dog hunted for the cached morsels. Jimmy never tried to fool the dog by caching inedible bits, but he did understand the dog's motivation. As the dog sniffed for caches, Jimmy would walk alongside and just when it seemed the dog would discover a prize, the crow would rush ahead and move the food to a more secure position.

Some crows may even comprehend when other species are mentally handicapped. Valerie Allmendinger was enjoying a short vacation at Harrison Hot Springs in British Columbia in the mid-1980s when she noticed the cook tossing whole slices of bread to a small flock of Canada geese. But the geese got none of the bread. Instead, a dozen crows assembled and quickly concealed the carbo-loaded bonanza from the geese. For twenty minutes, each bird picked up a large, red maple leaf and placed it exactly on top of a bread slice. As the geese headed toward the bread, they seemed confused at its rapid disappearance. Under these circumstances the hungry geese did not understand that the bread was still right in front of them, just under the leaves. They honked and milled about in search of the visible. For a goose, out of sight seems indeed to be out of mind. And for a crow this mental lapse is an opportunity to exploit.

The interesting premise that a crow might know what another being is (or is not) thinking is a mental attribute known only in people and our closest primate relatives. This complex cognitive property is called possessing a "theory of mind," and it requires being able to imagine and consider what others know. The anecdotes above are suggestive but far from conclusive. The Florida crows might have mistaken the clay they cached for actual fish, or perhaps there was some nutritional value in the clay that the crow required. Jimmy, however, would have little reason to move his food if he did not understand that doing so would foil the dog. Still, this is a rather simple association—dogs eat caches—that could be learned through a series of bad experiences. If only some dogs stole food, would Jimmy understand that all dogs don't lust for his caches in the same way? Controlled experiments on captive common ravens and western scrub-jays suggest that understanding the intent of different dogs would have been no problem for Jimmy the crow.

Common ravens swarm to large animal carcasses and quickly strip a once-meaty beast into a heap of bone and hide. The daily energy demands of a feasting raven, even the most subordinate bird that waits for others to eat first, is easily met during a few hours at the food bonanza. But often birds dally at the meat throughout the day, taking away far more than they can digest to hide this surplus in the snow or underground, carefully concealing it from the masses that share their meals. In a large aviary, Bernd Heinrich and his colleagues watched ravens eat and hide leftovers, and they documented just how aware ravens are of the knowledge of their peers. Ravens try to cache out of sight of their cagemates. They sneak away or hide behind a tree and look before they cache. They try to fake each other out, seeming to cache in one place but holding their food deep in their mouths and moving to cache it elsewhere. Sometimes they have to cache with others watching. When they do, and if that bird later approaches the cache location, the ever-watchful cache owner quickly flies ahead of the potential thief and reclaims the treat. Just like Jimmy and the dog. If another raven, one who was not present when the cache was made, approaches a cache site, the owner of the cache sits unfazed. Ravens know that some individuals know where they have hidden treasures and that other individuals do not possess this information, and they use this knowledge of others' minds to defend and move foods as little as possible.

Ravens also comprehend what people know. Jene Jernigan was living at the beach south of Half Moon Bay in California when she formed a strong bond with a wild raven, not unlike the bond we formed with Al in British Columbia. One day the raven recovered a cached white plastic snap cap from under a stone near Jene's feet. The raven seemed to show off the prize and then recached it. Jene dug up the cache, admired the cap, and returned it. The raven, perhaps fearing Jene would again pilfer his treasure, quickly removed it and recached it in a new, distant location.

Ravens may know what is on our minds because they pay close attention to our eyes. Like many social animals, ravens often look and explore in the direction that others are looking or facing. This tendency to follow another's gaze may be a simple response to an animal's orientation, or it may indicate that the observer understands what the staring animal knows. A set of experiments by Thomas Bugnyar and others suggests it is the latter. Professor Bugnyar reasoned that if ravens simply respond to a person's gaze as a general cue, then they might themselves look in that direction or follow along the gazed path for something of interest. But if a raven understands that a gaze implies a visual experience much like one's own experience, then a raven might follow a gaze not simply in a straight line, but even around barriers. Very young ravens were attentive to the gaze of people, but they looked only where people looked or walked in the direction of the gaze until they encountered a barrier. They did not seem to perceive that people might be looking past the barrier.

As ravens age, their understanding grows, similar to the way a toddler learns that a person who hides behind a door is still there, just out of sight. By the age of six months, the ravens understand that a person's gaze might indicate the direction to something important. The researchers' gaze might reveal the place where they ate or left interesting objects. As experimenters looked toward areas of the ravens' cage that were behind short walls, the ravens no longer stopped at the wall. Now they jumped atop it, flew around it, peered over and beyond it, and searched throughout the area at which people gazed. The acute attention that ravens pay to our subtle signals underscores the degree to which they can draw conclusions from our body language. They perceive our intentions even though we may not be consciously aware of them.

Crows also understand human intentions and seem to monitor and assess our motives. In the 1940s a class of students at the University of North Texas experimented with crows that pestered and were frequent targets of irate pecan farmers. The nine students built a large blind in the pecan orchard that resembled a haystack and gathered in the blind to watch the crows, only to learn that they, not the crows, were the objects of study. With students ensconced in the blind, the crows ate no pecans but flew to nearby trees to perch and remain alert. Frustrated, the students began leaving the blind in groups of random size: one, four, or eight students would leave the blind, and the others would record the actions of the birds. The crows, wary of people and able to count, did not resume their pecan-raiding behavior until eight students left.

Many animals have good numerical sense, and some, like the crows outside the blind, can keep track of hidden objects. Alex the

Ravens follow the gaze of people, here even around a barrier, suggesting that they know what the experimenter knows.

parrot and corvids understand numerical quantities up to about six; this seems the maximum number of objects that they can quickly assess without counting each individual item. The North Texas crows seemed genuinely to be counting, not just making quick assessments or estimates. One wonders why a crow would have any capacity to count at all until we consider that some crows and ravens lay up to eight eggs in a single clutch. Could they be keeping track of their eggs, nestlings, and fledglings by counting them?

The Texas crows were more than numerically gifted. They behaved in a way that suggested they knew that people were intent upon inflicting them with harm—but not all people. The students devised a clever twist to their experiment. As they marched out of the haystack, some carried nothing, but others carried either broomsticks or shotguns. Regardless of how many in the group carried guns or how the guns were held—over the shoulder or broken open across the forearm—the crows scattered at the sight of any armed student. But the crows uniformly ignored students with broomsticks. Crows know guns, but do they know people with guns are thinking about killing crows? We suspect they do, but we cannot be sure; the motivation of the person was not manipulated independently of the gun, and guns were never tested without a person present. It seems to us that crows understand the connection between people, guns, and harm, but we don't know if they understand the separate elements of this complex association.

Given that crows, ravens, and jays assume knowledgeable potential thieves will rob a cache, it seems more likely that crows also know that armed people—not just the guns they wield—are dangerous. Crows can quickly read human body language and determine an appropriate response to it. Should a crow detect a human's steady gaze or someone walking in its direction, it is quickly alerted and prepared to flee.

Since crows know what their fellow beings are thinking and planning, it is possible that they actively try to change others' minds. The possibility that crows knowingly manipulate people was raised by Gary Clark and many others who daily feed wild corvids, whom we visited during our research. Gary carefully trims the skin and fat from fresh chicken. A pound or more of the greasy mess bloats a plastic bag next to him. He grabs the swollen container, a few slices of old pizza, and a tub of dog kibble and peanuts and heads out to feed the crows. His collection of scraps and goodies today is usual crow fare. We've done our own share of direct recycling from kitchen to crow.

Today Gary is rewarded nearly immediately after he steps into his backyard, rattles the dry dog-food-and-peanut dinner "bell," and plops the pizza and chicken onto the elevated TV-dinner tray rooted by four legs into his lawn. Shortly after we hear a single crow call in the distance, a crow alights in the cherry tree above the feeder. Then two more arrive, and within a few minutes, fifteen to twenty-five hungry crows drop from a wet sky into a typical suburban backyard to devour the carefully blended feast. We are transfixed by the black shapes beyond the glass doors as they fly acrobatically, call, chase, and wrestle with one another, even jostling for tidbits with several resident squirrels. To Gary's amusement, the ever-changing scene is fascinating, engrossing, and yet mostly predictable. These wild animals take him from the confinement of a well-insulated, fully bric-a-brac'd house into a wild, dynamic scene. They also provide the additional benefit of lawn care as some of the crows pluck crane-fly larvae from his well-tailored grass.

Today's activity was pretty routine, but barely two years before, at this same spot, something very unusual happened. Gary had been heaping his tray with crow chow for a couple of years, when just after Valentine's Day 2006 he talked to a small hungry flock gathered above him in the cherry tree. A big man with a military bearing and haircut, he nevertheless evoked an open-armed solicitation and asked the assembled birds: *"Hey, how come you never bring me anything? I always give you food, and you never bring me anything."* With that he returned to the house, and the crows quickly descended to eat everything he had left on the tray. Late in the afternoon, long after the feeder had been picked clean, a small purple object on the feeding tray caught Gary's eye. *What the hell is that?* he asked himself. Upon inspection, he was stunned to see a candy heart centered on its surface. It was well worn, but clearly visible on one side was the word "love."

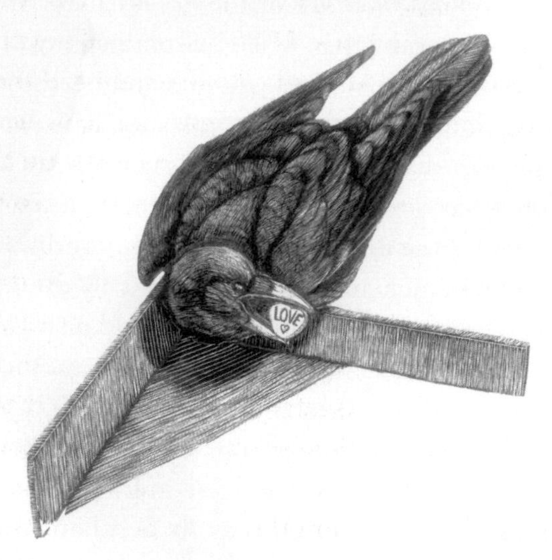

Crow leaving a gift on the feeder where good food is provided.

Right. You've got to be joking. A touching gift from beast to man were our first thoughts when Gary emailed us his story. We came up with seven hypotheses that are consistent with the episode.

1. Crows understand the spoken and written word.
2. Some person had pulled Gary's chain.
3. Gary had pulled our chains.
4. Some other nonhuman animal had carried the heart to the tray and dropped it by accident.
5. Gary had encountered a former pet or trained crow.
6. Gifting was real, but a mistake.
7. Gifting was real and purposeful, maybe a form of reciprocity or coercion by the crow.

How do these seven plausible, scientific hypotheses stand up to the data? Consider the hypothesis of chain pulling. We figured either Gary was crazy or the butt of an interesting prank. Gary lives only a few miles from us, so we paid him a visit. Gary's yard is fenced. He and his wife, Sue, have no children at home. Sue suffers from rheumatoid arthritis and, on crutches, wouldn't likely sneak out to the backyard merely for a prank. His neighbors do not appreciate the crows' daily ruckus, and if they could have accessed the feeder, certainly they would have left a different message. This heart, as unbelievable as it seems, was unlikely to have been left on the feeder by a person. It could have been left on the feeder by a crow or possibly a squirrel.

Two years later, Gary is still awed by the gift. And now he also has a small metal butterfly, a round cylinder of cement, a fir cone, a twig, and a dried berry to add to the collection. All are carefully kept in a special jar. Each item was left on his feeder after crows devoured what he provided. Gary and Sue are convinced the crows, or at least one stalwart sentinel crow who routinely patrolled the yard, understood Gary's plea and gifted him the candy heart. Of late, the gifting is on hold. Perhaps the generous, familiar crow no longer lives in the neighborhood.

Gary is not the first or only person to receive a gift from a crow.

Nancy from Bristol, Indiana, called in to the *Diane Rehm* radio talk show to tell us about a crow that had visited her and given her a small wooden bead. She had been sitting on pillows, reading a book in her yard, when suddenly a crow dropped from the sky, landed on her lap, and left an inch-long necklace bead on her leg. Nancy was shocked. She did not feed crows and in fact rarely noticed them in her neighborhood. She kept the bead. Nancy did not talk to her crows and there was no writing on the bead, so we can safely reject hypothesis 1. Crows need not understand the spoken or written word to leave gifts.

The bold behavior of Nancy's bird suggested a former pet animal, habituated or imprinted on humans. Raising a crow as a pet is a common albeit, today, illegal activity. Owners of pet crows report receiving bottle caps, pull tabs, coins, false teeth, feathers, and flowers. One pet crow even offered a prized slice of roast beef to an ailing canine companion. But there was more.

Leona, from Missouri, routinely receives shards of colored glass in her bird feeder—gifts from wild crows in exchange for sunflower seeds. Gayle had a red-and-white toy bomb dropped beside her by a wild crow. Molly LeMaster, who also talks softly to crows in her yard, has received a small bird wing, a frog leg, a steak bone, a marble, a shiny rock, and a bracelet charm. Barbara Arnold from Port Townsend, Washington, tenders a box of presents collected from the family of crows that routinely visited and fed in her backyard, including half of a red poker chip, a penny, a paper clip, a red die, bits of colored glass, a nail, a safety pin, colored rubber bands, bits of pottery, a blue glass bead, a red wire twist tie, a tiny pin that says Loyal Legion Week 1933, and, as Barbara quickly adds, the best thing of all—a blue plastic Cap'n Crunch figurine.

Beth, a crow feeder from Seattle, would regularly leave dog kibble for her crow flock by placing the food on the sidewalk as she strolled about her neighborhood. There was no clamor or competition for the food as each bird would descend separately to pick up its share as she moved along. On one occasion, however, her routine took an unusual turn. Hearing the sound of metal hitting the hard cement behind her, she turned and saw that one of the crows had dropped what seemed to be a bright house key just to the side of the tiny stack of food. The crow took the treat and left the key for Beth. Clearly, the sheer number of independent observations of gifting crows allows us to safely reject the first five hypotheses and concentrate on deciphering whether gifting is intentional or a simple mistake.

Sometimes when crows follow a person who is feeding them, they drop gifts, such as this key.

Sometimes people are only indirect recipients of crow gifts. Janet Brandt has a decorative iron crow in her yard under which a local wild crow stashed a dead sparrow nestling—right under the iron crow's foot. Relatives of crows also have been observed leaving gifts. In Kuna, Idaho, Eric Freeman leaves dead mice that he's trapped in his barn on a barrel for magpies, close relatives of crows. The magpies in turn leave shiny objects on the barrel.

There is little in the scientific literature about gifting behavior. Dolphins are known to throw fish to birds, and they occasionally leave fish for people with whom they closely interact. A variety of birds gather the sorts of objects crows use as gifts. California condors, for example, deliver bottle caps, pull tabs, and bits of plastic to their nestlings, which they unfortunately eat, and which kills them. Many raptors decorate their nests with fresh greenery, perhaps serving a sanitary or insecticidal role. Ravens often incorporate brightly colored string or wire artifacts into their bulky nests. Evon Zerbetz even reports one industrious pair interlacing welding rods into a nest and feathering the unique structure with Teflon tape, work gloves, screwdrivers, and brightly colored flagging. Male bowerbirds obsessively collect shiny rocks, glass, and plastic to pave their display courts and impress potential mates. And many members of the crow tribe are well known to collect and cache shiny objects. So birds pick up interesting objects, but no scientist has yet reported the connection between a penchant for gathering with a habit of giving to people.

Crows routinely gather and store bright items, so some of these gifts could be simple, unintentional by-products of such an inclination. A crow carrying a shiny object spots an important opportunity to feed, lands on the feeder, drops the item, and because its mouth is full, it cannot regain what it dropped. Opting for the feeding opportunity, it leaves the prize. Gifts placed by accident should often be irrelevant to those receiving the gift, but most people who reported obtaining gifts from crows regularly feed crows. And many of the gifts they obtain were of human origin. On the other hand, people like Nancy, who do not feed crows, also get gifts, and many gifts are natural products like stones, twigs, feathers, and flowers. The evidence is not compelling enough to reject hypothesis 6, that gifting could be accidental. But if a crow values an accidental gift before eating the food it found, why wouldn't the crow retrieve that object after eating? Why weren't gifts stolen by other crows after the crow fed or cached quickly nearby before eating? These nagging questions compelled us to dig deeper into the possibility that crows purposefully leave gifts.

The gift of a nestling to Janet's iron crow suggests that some offerings may be intentional. Her crow was apparently duped into leaving food for a potential mate or social partner. Could crows at feeders be leaving anthropogenic gifts to impress people, or perhaps other crows? If gifts were intended for crows, we would expect them to be carried off by the recipient. Maybe some are. But the possibility exists that gifts left behind are intended to court or impress people important in a crow's life. This notion gains some support when we learn that Gary stopped getting gifts about the time he also noticed that a particular crow, who had been regularly present in his yard, had vanished. Perhaps that crow had been paying homage to another species that shares its territory and provides regular and dependable sustenance.

In Arizona, at least one crow seems to have directed his gifts toward an important person. Ornithologist Russell Balda and his wife, Judith, were startled one early summer morning to see a crow hanging by its feet on the wooden slats of their fence. They rushed out to free the bird, wrapped it in a calming towel and set it on a rock wall under a large oak tree. The stunned and exhausted bird lay limp for nearly twenty minutes as Judith talked softly to it. A robin, seeing the crow as a threat, dived at the bird, which was jolted back to life and flew off, the irate robin in close pursuit. Later that autumn Judith noticed a crow hanging about the yard and visiting the bird feeder on her deck. Again she talked softly to the bird, and soon gifts started to appear: a dead mouse, regurgitated bits of seed and meat, and sticks. Judith saw the crow leave some items, but others just appeared after the bird had been around. This bird directed its actions at Judith in the place where a crow's life had been restored. It seems certain that this was the bird Judith rescued, and it also seems that this crow had a crush on Judith. The crow's actions were not unlike those of a bird courting a potential mate or one repaying a remembered debt to another being.

We are left with two hypotheses that are consistent with the data: gifting is real, and it may be accidental or intentional. Natural selection could favor purposeful gifting behavior in crows. Gift-giving crows are rewarded by people with protection and food, which improve their survival and reproductive success. Mentally, crows would have no trouble associating a gift with food. Perhaps more accounts or controlled experiments will tilt the weight of evidence in favor of either the accidental or intentional possibility.

Reciprocity may not be a practice exclusive to humans. The ability to quickly associate behavior with reward that is so prevalent in corvids may underlie their innovative gifting behavior. Leaving gifts suggests that crows understand the benefit of reciprocating past acts that have benefited them and also that they anticipate future reward. In either case, purposeful gifting is more than a simple response to a food stimulus. It is a planned activity; the crow has to plan to bring the gift and plan to leave the gift. Because crows do not retrieve their gifts, they must have a sense of purpose, and they must plan for the future. Conceiving the future requires substantial cognitive ability, including mental time traveling and insight. It seems entirely possible that scrub-jays, ravens, New Caledonian crows, and rooks have these mental abilities.

Gifting interactions certainly strengthen people's connection to crows and their relatives. The salient, persistent, and reciprocal nature of these personal interactions may be important to our cultural evolution. In today's society, bird feeding is a growing hobby, a culture reinforced by the delight that birds offer and the sometimes fascinating conundrums they pose.

Other stories and legends indicate the historical depth of this influence. The Old Testament tells of ravens sustaining the prophet Elijah with gifts of food as he sought retreat in the desert. Ben Jonson wrote in his classic play from the 1600s, *Volpone,* of a fox feigning death who is visited by Corvino, the merchant crow. In gratitude and respect, Corvino gives Volpone a pearl and a diamond. Perhaps Mr. Jonson or one of his friends had been gifted by one of London's many crows, magpies, jackdaws, or ravens.

Among the most remarkable myths concerning crows and ravens are those from the native peoples of the Pacific Northwest. For thousands of years, Kwakiutl, Haida, and Tlingit shamans were deeply moved by their interactions with these birds. Native people routinely spoke softly to animals they encountered to calm them and thank them for giving their lives in the hunt or harvest. They were necessarily careful observers of nature, using animal signals to find foods. Their religion centered on the Creator Raven and glorified the gifts He brought them: light, salmon, caribou, and bear. Their ordered world itself was a gift from Raven. And so, too, the lives of modern ordinary people like Gary are brightened by special, but perhaps not so unusual, gifts from crows.

Crow gifts may be much like human gifts, cultural traditions that have developed to bind important partners by reciprocity. Crows benefitted from human reverence and generosity for thousands of years, and natural selection could easily favor a culture of interest in human artifacts and use of those artifacts by crows to gain our favor. Whether crows are conscious of their actions, simply exploiting us, or more deeply thinking about us does not matter. Gifting is an effective strategy for a noisy, brash, and often reviled pest of a bird to charm its way into our society. And this remarkable charm invites us to contemplate and reflect on the common ground we share. This connection continues. Last year Lois Magnusson of Seattle could hardly wait to tell us about her neighbor's crows, who are routinely fed on a white paper plate. One crow, when hungry, now takes the empty plate and carefully puts it by the back door. Gifting. Begging. Demanding. It's a familiar pattern of behavior.

Crows are clearly clever problem solvers, routinely making associations, even planning and using insight to succeed in a world constantly reshuffled by people. They think about what we are doing and exploit us when it suits their needs: filching food we leave for others, using our engineered contrivances for their own good, planning hunting and fishing expeditions. They even prime the pump of our generosity with small gifts. They are a common and persistent entry to encountering nature and a reminder that we are an inseparable part of it. They provoke us to a point where, like them or not, we must notice and consider them. Their antics mystify and entertain, and their physical origins and designs provide insights to our own evolutionary trajectory. Becoming something of an essential part of our lives may be their greatest survival tool of all.

Ravens surfing the Colorado wind.

6

Frolic

A HEFTY WIND assaulted the cliff wall to produce a powerful late afternoon updraft. Above a scenic viewpoint in Colorado's Rocky Mountain National Park, geysers of air call like sirens to ravens, as big waves beckon to surfers. Neither ravens nor humans can resist the challenge, fun, and celebration that is powered by nature's energetic breath. Ravens dive, dip, chase, roll, tumble, somersault, and shout as they ride the wind. The stronger the wind the better. Rarely, if ever, do they hunker down and wait for the calm.

On this windy October day in 1999, a group of eight ravens took their practice to a more sophisticated level and surfed the gale, gripping quarter-inch thin, concave arcs of tree bark in their feet. Pairs of birds spread their wings and launched into the air. Without flapping, the birds used their legs to angle their inch-and-a-half-wide by six-inch-long bark surfboards into the wind so that they soared, then dived, slipped this way and then that. Each bird hung all eight of its toes off its board and maneuvered it perpendicular to its body, like a snowboarder slows a too-rapid descent. The surfers rose to fifty feet above the overlook, rarely higher. Oblivious of the human onlookers below, one pair surfed while the others rallied and chased, trying to steal the boards and take their turns at the fun.

For an hour the games mesmerized Emilie and George Rankin. Then the birds flew en masse down the mountain, to the valley, and likely to a safe night roost. Observing the ravens' surfing party was a highlight of the Rankins' autumn trip and remained an indelible memory a decade and a half later.

Frivolous antics such as the windsurfing party are routine among ravens. But can we really call such activity "play"? Absolutely! A few decades ago, most scientists scoffed at the notion of animals playing, especially those other than monkeys, chimps, and apes. But today the scientific study of play is robust and concerns the origins, functions, evolution, and neural basis of play in people, turtles, octopuses, and insects.

The most difficult aspect of studying play is actually defining play. We know play when we see it, especially among our own species or our close domestic associates, but it is a challenge to objectively sort purposeful behavior from play by creatures whose signals for initiating play we do not understand. The surfing ravens might actually be participating in ritualized courtship or status displays, or their flights could be signaling newfound foods and rallying one another to roost, which are known functions of soaring. Nonetheless, we are confident that the surfers were playing.

Ethology—the study of animal behavior—gives us five criteria for considering a behavior to be playful. Play (1) includes actions not immediately needed to survive; (2) is voluntary, pleasurable, or self-rewarding; (3) includes typical actions done in incomplete, exaggerated, awkward, or novel ways or not all the time; (4) is often repeated in similar, but not necessarily rigidly stereotyped performances; and (5) occurs in benign conditions such as when the players are well fed, healthy, and free from stress. Surfing ravens meet each of these requirements.

An extremely complex form of play, surfing is technically a mix of "social," "locomotor," and "object" play. Crows in Seattle also combined these modes of play as they flew above the end zone at a University of Washington football game. Just as the football players used motion, objects, and one another to play, so, too, did a flock of about fifty crows. One crow had a ball of paper that it carried with its feet for a few flaps before dropping it. The other crows rushed to grab the fumble. One crow quickly footed the paper midair and flew on, just ahead of the loud mob. Rising high, the crow dropped the paper and another bird recovered it. This whole sequence replayed over and over high above the stadium and, given the poor performance by our team on the field that day, was a welcome diversion for the fans.

Many birds play. In a third of all orders of birds, voluntary, novel, immediately unnecessary, repeated, stress-free movements, interactions with objects, or games among individuals have been recorded. Chickens, ducks, seabirds, cranes, woodpeckers, owls, hornbills, swifts, turacos, parrots, pigeons, songbirds, hawks, and cuckoos play. But no group of birds has been reported to play as frequently, as variably, or with as much complexity as the corvids.

In Japan, jungle crows hurl themselves headfirst into updrafts deflected by high-rise buildings. The ebony missiles plunge into the wind before snapping open their wings for a fast ride back up to the roof. This social game involves many crows who take turns hurling, riding, kiting, and gabbing on the building roof. Often one bird claims a prominent structure like an antenna or lightning rod as a launch site. Others take turns, even cuing up to jump from the prominence. Competition for launch sites sets off side games of "king of the mountain" as waiting crows try to displace the one atop the spire.

One late November afternoon, Tony witnessed an episode of social wind riding by crows that illustrates the unnecessary, voluntary, novel, repeated, and relaxed criteria that define play. He was driving north from Seattle as the cloud cover broke and sunshine brought unexpected warmth to the day. Along the horizon he could see a long line of hundreds of crows flying east toward their roost in the town of Kenmore. When he stopped for gas, the crows, which he then judged to be in the thousands, were overhead and still flying by in great numbers. But they'd begun to do something unusual. Many of the birds were breaking away from the flock and diving down to swirl and float over the top of an enormous water tower. They randomly chased one another over the curve of the dome top, occasionally parachuting down to perch there. The afternoon sun had heated the bright white metal skin of the water tower's top sufficiently to create a shell of rising warm air. The birds atop the tower leaped into the thermal shell and with little effort rose quickly up over the tower to join the gathering numbers of crows already suspended there. Looking like irregular black balloons with outstretched wings, the birds rose and fell effortlessly in the updrafts. For a quarter of an hour crows continued to break away from the larger, commuting flock to join in the activity until a hundred or more hovered, tumbled, and chased within the warm air over the water tower. As more birds flew into the frolic, others would break away from the top of the hovering mass of birds and fall back into the line headed for the roost.

Crows breaking from flight path to play in warm air rising from a water tower.

Playing corvids often careen down slippery slopes. In the mountains of Japan, Alaska, Canada, and Wales, ravens have been seen and photographed sliding like otters down icy, pitched roofs, and steep, snowy banks. They slide headfirst on their bellies and backs or roll sideways, like kids in a barrel. Sometimes they grab a bite of snow or snowball and carry it as they swoosh. When friction finally ends the ride, they lope up the slope to repeat the fun. In Russia, hooded crows have been filmed using plastic lids as sleds to slide down steep roofs, while others use their bodies to slide down the gilded cupolas of historic Russian Orthodox churches, which polishes the bright domes. Our Seattle crows can only dream about such golden slopes, but they do slide down rain-slickened roofs of metal and cedar.

Ravens often slide in snow.

When ravens and crows can't slide, they ride. Gripping branches or wires with their beaks or feet, many corvids hang and bounce, often upside down. Mary Palm from Lake Oswego, Oregon, watched a crow sitting on a springy, dead branch silently bob its head up and down to start the branch flexing. Up and down rode the crow. When the swinging stopped, the crow started bobbing again to restart the amusement ride. Donna Winter's adopted crow frequently rode a rotating sprinkler, much as children ride a merry-go-round, except the bird got a thrill and a bath simultaneously.

Ravens hang from springy branches and also ride wild wires in gatherings resembling rodeos. Richard Borgen watched as a group of ravens in eastern Canada tried to grab a slack wire whipped up by strong wind. One bird succeeded in gripping the wire with its foot and hung on, whipsawed by the motion, for twenty tumultuous seconds. The other ravens watched and tried to grab a ride for themselves as soon as one bird was thrown from the bucking wire. Japanese jungle crows are even more acrobatic. One windy day when several crows were perched on a power line, one bird plunged forward and swung around the line like a gymnast doing a high-bar routine. The other crows hung on and watched but did not swing.

Corvids also play more traditional games. John and his wife, Colleen, spent three years living with ravens in the mountains of western Maine. With Bernd Heinrich, the scientist who challenged ravens with string tests, they raised, watched, and followed several hundred captive and wild ravens as they searched out and fed on the carcasses of large animals, roosted in nighttime congregations, developed their voices, secured their lifelong mates, and raised their ravenous broods. We researchers were imbedded in the intimacies of raven society, of which play was a common feature. The young birds were the most playful. Daily, they engaged in games of tug-o-war and keep-away. Any scrap of food, chunk of wood, nut, leaf, or pebble served as a toy.

A typical play bout unfolded as follows. One bird carried or picked at a twig as a second bird approached, typically fluffing its feathers in a nonthreatening posture. The invitations to play were not as obvious to us as were the play bows of our dogs, but the other ravens certainly recognized and distinguished the intent to play from the intent to challenge.

A bird seeking to play triggered one of two reactions from another bird in possession of a potential toy. If the toy was amenable to a tug—a long stick, strand of sinew, or bone—the possessor typically lay down and rolled onto its side or back and allowed the approaching bird to beak the object. Then the games of strength began. Both birds pulled in opposite directions, dragging each other growling through the snow or across the forest floor. As soon as one bird let loose, the other stopped and offered the toy for a rematch. The pulling restarted. Winners and losers frequently shifted roles, and only rarely did play escalate to aggression, which otherwise is frequent among assembled ravens. Hardly a day passed when birds aged a few months to a few years old did not play tug-o-war.

Less frequently, a bird carrying a rock, hunk of meat, or bit of woody debris ran or flew with its prize as another approached. This usually set off a wild chase involving several birds. As the bird with the object dodged its pursuers, maneuvering like a fighter pilot intent on evading an enemy, the prize was often dropped, picked up by another, and the chase restarted in a new direction. Back and forth the frolicking ravens rocked through their world, momentum shifting this way and that. These activities were spontaneous romps by juveniles in relaxed situations—typically after a winter feast when predators were unseen and the activity had no apparent survival benefits. The birds often used inedible items and repeated their actions in unpredictable but similar fashion.

Young ravens often play tug-o-war.

Pet corvids play with all sorts of toys. A young raven in the Copenhagen Zoological Gardens played catch with pebbles, snail shells, and a rubber ball. This bird often lay on its back and tossed its toys between foot and beak or vertically high into the air. It rarely missed.

Esther Woolfson's rook, whom she called Chicken, played with colorful, rubber mice. Chicken threw the mice about by their elastic tails, pulled and pounced at them in feigned predatory attacks. Woolfson's magpie, Spike, went even further. He played ball alone or with whomever would join the fun. Spike fetched balls thrown by the Woolfson family and begged for the game to continue, just as a faithful dog would do.

A captive raven at the Copenhagen Zoo often played with a ball.

While some corvid attitude is difficult for people to perceive, there are universally recognized aspects of playful posture, voice, and movement. Because of this, crows can play with other species. Carole Anne Coffey took in an injured nestling crow she named Omar and raised it as a full family member for almost eight years now. Omar plays with newspaper—often shredding it, soaking it in his water bowl, and draping it from his perch to improvise a shower. He also plays with the Coffeys' cat. Omar grips the end of a shoelace and pulls it along the floor in front of the cat until the feline pounces, tugs, and scurries after the toy. The two pets, enemies in nearly any other setting, clearly recognize each other's intentions and play. The raven from the Copenhagen Zoological Gardens frequently engaged a resident dog in ring-around-the-rosy. Raven and dog ran to a tree, circled the trunk, and countered each other's moves to keep the tree between them as they chased round and round. The reports don't mention cawing and barking, but we can hardly imagine either playful animal was quiet.

Omar, a pet crow, lured the family cat into games with a shoelace.

In the wild, crows and ravens interact with other species in ways that suggest play. On Kinkazan Island in northern Japan, jungle crows pick up deer feces—dry pellets of dung—and deftly wedge them in the deer's ears. Lawrence Kilham observed crows and turkeys chasing each other as he scattered corn. Several of the crows teased the turkeys by luring them with beaks full of maize, flying low over and in front of them, and even pouncing upon their backs. The turkeys gave chase; both species went back and forth. The two avians may have been simply competing, but they clearly were communicating.

Tony watched a raven spend nearly an hour pestering a young wild turkey unable to clear a wire fence and rejoin its flock. Each time the big game bird stuck its head through the mesh fence and struggled to pull its enormous girth through the narrow gap, the raven hopped over and pulled hard on the turkey's tail. Perhaps the turkey's tail was plucked to steal a meal or wear down an oversized prey, but we think the raven plucked the turkey's tail for entertainment. Reports of pet corvids yanking the tail of the dog or cat with which they reside are commonplace as we showed in an earlier chapter. The response of the wild turkey, like that of the domestic cat or dog, is exciting and striking, and perhaps this is enough motive for continuing.

Raven pulling the tail of a turkey.

Playful birds may be vulnerable to injury or surprise attack, and play seems to waste energy needed to survive, so we wonder why natural selection does not weed out those that play. How might a playful bird benefit later in life from its experience? The search for the reasons birds play, both immediately and ultimately, takes us down three interrelated paths. The first regards features of the corvid lifestyle that promote play. The second considers the functions of play, many of which simultaneously drive the player. The third addresses the notion that play is fun by exploring how playing floods the crow brain with motivating, regulating, and pleasurable chemicals.

The corvine existence enables and demands play. Longevity—many corvids live for decades in the wild—allows a protracted juvenile period relative to most birds wherein a young bird has time to experience its physical and biological world. A young robin just out of the nest might spend a week with its parents and a winter with a large flock of peers and older birds before attempting to stake a breeding claim as a yearling. A young jay, magpie, crow, or raven will spend months with its parents and siblings on familiar ground, and most will live years before breeding.

A long life before breeding is typically an extremely social period for a young corvid. In some species, young birds remain on their homeland as apprentices, helping their parents raise subsequent broods, defending territory, and acting as sentinels on guard for danger. In other species, juveniles disperse after a few months to drift in and out of large, dynamic social aggregations and intimate partnerships before forming a lifelong pair bond and entering the adult world. All the while young corvids are exposed to a rich diversity of foods and situations. Their generalist lifestyle brings them into contact with fruits, seeds, insects, mice, lizards, snakes, frogs, eggs, trash, and carrion of all types that they test and appraise before eating or ignoring. A period of months to years in often relaxed, familiar, even protected settings that provide many new and ever-changing social and dining situations affords a young corvid the opportunity to play with peers and objects. Play is an active learning strategy that continues throughout its lifetime to inform the bird about the complexities of its world.

The activity of play builds and refines synapses in the bird's brain. In predators, play allows a young animal to perfect hunting, killing, and social skills without risk of injury. In corvids, whose weapons are relatively diminutive, such learning may be of secondary importance to the basic need to learn about potential foods, social companions, enemies, and survival strategies. Chasing a peer who displays a novel object may teach a young bird about a new food. A bird in pursuit learns how to read and respond to social signals, which can advertise a companion's power, status, reproductive readiness, and motivation to compete or help. Responding to an invitation to tug a stick may allow a young raven to take the measure of a potential rival and learn how to engage a social partner and read another's postural signals. Pinching the tail of a turkey, meerkat, wolf, or dog answers questions: Is this food or a deadly force? If deadly, is it satiated or coiled for the next attack? Poking a larger beast might perfect a young crow's strategies that he will need later to pilfer from a predator or hunt cooperatively to down large game. Ravens, for example, exploit wolves that kill, open up, and provide carcasses upon which to scavenge. There is a problem, however, as the power to kill an elk can also kill a raven. Learning how a potential predator reacts and being able to measure critical distances are important survival skills.

John once approached a raven in Gardiner, Montana, whose injured wing prevented flight. The shuffling bird drew him into a prolonged bout of ring-around-the-rosy. The bird dashed to the base of a lone pine tree. As John went left, the raven circled to its left. John's dash to the right was countered by a quick raven hop to its right. And so it went, the raven moving precisely enough to keep the large tree trunk between them. Perhaps this bird first learned the technique it now used against John as a playful fledgling, just as the raven from Copenhagen learned to elude a familiar but dangerous dog. Playing corvids innovate, practice, and perfect skills and manners that are necessary adaptations to flexible, social lifestyles. By testing their world and sensing the response, a playing corvid is shaping its developing neural circuits: forming new synapses, strengthening useful ones, and losing others that are unnecessary. The young corvid's brain creates expectation to which it will compare new experiences.

Play builds memories that allow an older bird to live and breed better than a bird that does not play. This is what scientists call the "ultimate" value of a behavior: natural selection favors playful young animals because they survive to reproduce later in life. This force causes play to evolve in some species and not in others. The benefits of play must outweigh the costs of play—the exposure to stronger peers, dangerous predators, and poisons—in order to persist in a population. Natural selection shapes variation in playful actions based on the costs and benefits to survival and reproduction. But the actual motivation and gyration of a player responds to more immediate factors. These cues and stimulations to play—the "proximate" reasons why an animal plays—are fully under the control of the brain. Crows play ultimately because natural selection favors players, but in a very direct way they also play because the brain of a playful crow rewards the bird with pleasurable, comforting chemicals. In other words, without invoking conscious comprehension, crows play for the same reason a child plays—because it is fun!

"Fun" is not an abstract concept, nor is it something only humans can experience. Rather, we consider "fun" to be the feeling produced by chemicals in the brain that reward certain behaviors. Eating a tasty meal, drinking a sweet liquid, nuzzling a parent, having sex, and playing are some of the behaviors that are rewarded by the brains of vertebrates. When two ravens tug at a stick, when a crow parachutes with members of its flock in a stiff updraft, or when a raven slides down a sledding hill, the diminutive hypothalamus of their midbrains releases natural morphinelike substances. These opioids, especially beta-endorphin, bind to special receptors on neurons in other regions of the brain and in doing so produce euphoric feelings. Medical science uses this same natural reward system to dull our pain with morphine, which binds to these receptors. Substance abusers become addicted to the heightened or prolonged experience of these states.

The neurobiology of play, which is viewed as one of many rewarded behaviors, has been studied extensively in mammals. We know how the brain chemistry of lab rats varies as they wrestle in playful bouts, and birds and mammals, indeed likely all vertebrates, share the basic structures and chemicals involved in brain reward circuits. So it is reasonable to use the science of play to better understand what motivates, rewards, and choreographs a playful crow.

A connected circuit of brain regions, which some scientists refer to as the "social brain network," senses, interprets, and shapes complex interactions, like play, between animals. As an animal encounters others in its society—mates, offspring, adversaries, aids, familiars, strangers, and the like—each brain region responds with a distinct perspective: emotional, geographical, cost to benefit, expectation to realization, and historical. This multifaceted consideration orients an animal to its social situation. As electrical and chemical information from distant parts of the brain flows among the wires, or neurons, that connect the regions, or nodes, of the network, the brain senses and integrates it, then translates it into movements and vocalizations—an animal's behavior. The pattern of information tailors a bird's emotions—aggression, appeasement, fear, courtship, care, affiliation, and play—to its social setting.

Important regions in the social brain network include several areas in the basal forebrain: the amygdala, which acquires, retains, and shapes lasting memories of pleasant and unpleasant emotional experiences; the septum, which integrates information from the amygdala and hippocampus to affect social bonding and affiliative behavior; the preoptic area, which produces and releases steroids in response to stress and so adjusts the firing of neurons in the amygdala and pituitary gland to orchestrate flight or escalate fight and learning in fearful settings; and the hypothalamus, which produces the opioids that reward action with pleasure (see page 230 in the Appendix for an illustration of the social brain).

These brain areas are mostly buried deep in the cerebrum and are best seen from a cross-sectional view of the bird's brain. They do not act in isolation, of course. It seems the more we understand about the bird's brain, the more important the connections among distinct regions become. The social brain network also includes the integrative centers of the forebrain, notably the hippocampus and the executive center. Several regions in the midbrain affect muscular control, neural connections, and the formation of memories and associations by providing dopamine into the network. When these regions are stimulated in humans and birds, by other chemicals in the body like sex hormones, their action potentials cause a release of dopamine into the synapse. Many neurons fire in bursts, releasing pulses of dopamine into the striatum and forebrain, which has substantial, varied effects on an animal's behavior, many of which are hotly debated and not fully understood.

The obscure nucleus accumbens, a small region in the lower part of the forebrain not traditionally associated with complex cognitive function, is a critical junction box of the social brain network. Its name means "to lean against the septum," but its function is far more important than its dependent name indicates. For people, the nucleus accumbens has been implicated in underlying our addictive behavior as well as our emotional reactions to music. For the crow, it also likely underlies pleasurable activities, such as play.

A common, but *incorrect,* notion is that dopamine is the brain's "pleasure chemical." Opioids, not dopamine, reward behavior with pleasure. Dopamine, especially through its effects in the nucleus accumbens, helps regulate the flow of information among the many nodes of the social brain network. Dopamine profoundly influences learning and memory. As dopamine enhances the impact of a neuron's information on other neurons in the brain, it increases coordination and sequencing of movements, focuses the learner, and especially increases the motivation to seek reward. Because of these interrelated functions, decreasing levels of dopamine also decreases play.

Here is how this might work. When dopamine binds to a neuron, it facilitates the release of the neurotransmitter glutamate into the synapse, which makes nearby neurons more likely to ignite and relay an electric signal. With dopamine, signals from the amygdala, the septum, the hippocampus and the executive centers of the forebrain whiz through the nucleus accumbens and on to the motor-control regions where they enable the crow to coordinate its movements to the changing situation of a playful peer or bouncing ball. This is the performance-enhancing effect of dopamine. The electrical codes that cue a behavior might also loop back to the forebrain through the thalamus for reconsideration. As dopamine eases this looping, it builds motivation; the expectation of an action is compared with the actual outcome as the animal thinks about what it did, anticipates an outcome, and is driven to perform. The match in an animal's brain between expectation and performance provides feedback to the flow of dopamine and, as a result, the continued motivation to play and to learn new moves. Meeting expectation keeps dopamine levels at the status quo. Exceeding expectations increase, while unmet expectations dampen, dopamine release.

Let's go back to the flock of crows playing over the University of Washington football game. The crow carrying the paper ball among the flock is motivated to do so by the dopamine flooding its forebrain. The stimulus of the paper, the wind, and the other crows trigger neural connections primed to easily fire by dopamine. The ball is dropped, and the crow dives in anticipation of a catch. She has done this before, and the electrical code that coordinates a diving flight with a slight tilt of the head to make a catch with her beak is driving muscles and looping through her forebrain. A typical grab keeps dopamine levels up, and she is motivated to continue playing. The connections among neurons that enabled the catch are reinforced, and her behavior remains the same. Again she drops the ball, anticipating a catch. Her dive is embellished with a quick wing retraction, a roll to the left, and a grab with her feet. Success beyond expectation increases dopamine, motivating her to play on and strengthening—easing their ability to refire—new connections among the neurons that allowed the barrel roll. Amped up, she drops the ball a third time and executes a revised roll toward it. A miss reduces dopamine, lowers her motivation to continue, and provides a teaching moment. The neural connections that produced the miss are not reinforced because as dopamine declines so, too, does glutamate. Without glutamate, connected neurons do not communicate, and their synapses—those that produced a bad roll—wither. Connections among parts of the social brain are reshaped by experience. This is learning.

As associations between cues and pleasurable rewards are formed, the cues themselves begin to stimulate the release of dopamine, and this elicits desire to attain the actual reward. A dog jumping at the door when its master picks up its leash is motivated by dopamine released in response to seeing a reliable cue (the leash) to a reward (the release of endorphins pursuant to a walk). The sight of a toy, a tempting updraft, or a teasing peer might trigger the release of dopamine in a crow that has played with these items in the past. Dopamine builds motivation by aiding the learning of associations between environmental cues and rewards, but in the nucleus accumbens it also seems to affect motivation in a more subtle way. Animals with inadequate dopamine in the nucleus accumbens often are unwilling to expend the extra effort required to obtain preferred rewards, and they take easier routes to lesser rewards. Rats with low accumbens dopamine, for example, will eat available but not preferred lab chow, while those with high accumbens dopamine will shun ordinary chow and instead seek out a lever that requires them to work hard (push repeatedly) for preferred high-quality pellets. Because of dopamine in their nucleus accumbens, ravens may be motivated to take the extra time or flap their wings a bit harder to obtain the reward of social play, soaring for hours far from their roosting locations.

While dopamine increases the performance and permanence of neural circuits, its actions are critically influenced by the pleasurable rewards of endorphins. Both of these chemicals influence social behaviors like play by binding to neurons—often the same ones—in the social brain network. In birds, important binding sites for opioids like beta-endorphin are scattered throughout the forebrain and midbrain, including the nucleus accumbens, hippocampus, and amygdala. To visualize these areas, we invite you to revisit page 230 in the Appendix for illustrations of where these sites occur in a bird's brain.

When opioids bind, they typically inhibit neurotransmitter release, dampening neuron firing. This is why pain is deadened when endorphins bind to receptors near synapses in circuits from pain receptors. Two distinct opioid receptors are important to play, μ-receptors and κ-receptors. Beta-endorphins readily bind to μ-receptors that are prevalent throughout the social network of a bird's brain, and this blocks synapses that are important to agitated behaviors such as fleeing, fighting, and distress calling. Instead of fear or aggression, the results of endorphin binding—especially to μ-receptors in the thalamus and amygdala—are soothing social comfort, euphoria, and increased play. In contrast, when opioids bind to κ-receptors, which are also found throughout the social network of a bird's brain, they produce feelings of sadness and reduce play. The level of opioids in the brain during play determines the types of receptors they bind to and the way a bout of play unfolds. When opioids are moderately abundant, most bind to μ-receptors and play is vigorous, but when opioids are superabundant, many bind to κ-receptors and throttle down play. Reducing the intensity of play may have a survival value as it allows the individual to become more cautious and alert to the possibility of an impending threat.

The euphoria that crows and we humans feel when playing comes from opioids binding to the complex synapses of the nucleus accumbens. The more an animal plays, the more opioid reward it receives. Some of these endorphins bind to the midbrain, increasing the release of dopamine to the nucleus accumbens. The occurrence of dopamine and opioids in the nucleus accumbens may also be important to integrated learning. The presence of these chemicals makes it more likely that response to a location, a reward, and emotion can combine into a new and simpler neural course of action. Simplification saves energy and confusion; the tens of thousands of synapses converging on striatal neurons are reconstituted in fewer than one hundred synapses between these neurons and those in the midbrain.

The chemical rewards to a playful crow's brain form a complex brew that acts a bit like a low dose of pot and heroin. In addition to circulating opioids, other natural products of the brain, endocannabinoids, are also synthesized in some neurons and released into synapses where they affect play. Where they bind and whether they block excitatory or inhibitory neurochemicals determines their influence. If they bind in parts of the brain such as the hippocampus or cerebellum, their effects can disrupt memory formation and movement, respectively. But when they bind in other parts, including the nucleus accumbens and the ventral tegmental area, they remove inhibition to dopamine release and increase play. In rats, cannabinoids specifically increase the responsiveness of an animal to a playful partner. Cannabinoids are also found in plants, notably marijuana and chocolate. No wonder a raven will toss a ball for hours and surf the wild wind.

Stress causes another brew of chemicals to be released and affect behavior. In stressful situations, a vertebrate animal's adrenal glands produce corticosterone. This hormone binds to neurons in many parts of the brain and affects the production of enzymes and neurotransmitters, which in turn affect an animal's behavior by regulating production of other hormones, movement of muscles, and the formation of memories, especially ones about dangerous situations. When a crow or person plays, corticosterone binds to neurons in the brain stem, which increases the release of serotonin and norepinephrine into the social brain network. But when people are stressed, both of these chemicals typically increase antisocial behavior, focus us to recoil from a dangerous situation, and reduce the inclination to play. In low-stress environments, norepinephrine may actually stimulate play by focusing the player's attention on the toy or social partner rather than on other environmental distractions. This highlights the importance of the environment to natural play behavior. When foods become limited, predators lurk nearby, or humans provide disturbance, serotonin and norepinephrine increase in response to stress, and play is diminished.

Our surfing ravens likely had low serotonin levels and were pumped on dopamine, endorphin, and norepinephrine as they dipped and weaved on the air currents. These chemicals motivated, rewarded, and shaped their play, just as they affect other animals' activities, making them possible and fun.

A complex cascade of chemical reactions determines each move a player makes, but the pattern of behavior is further shaped as a whole. The relative stimulation of each region in the social brain network and the strength of their interconnected synapses can be thought of as forming distinct, neural "ensembles" of nodes wired together that code in the brain a complex behavior—the way an animal moves and what it expects—in response to a situation. The ensembles are shaped by learning; they're remembered and easily recalled under appropriate conditions to produce a wide range of actions such as play, care giving, aggression, and escape.

For our ravens, the places in their social brain network that responded to the thrill of surfing could form a unique ensemble. These brain regions and the strong connections among them might be triggered again when the conditions are right for the feathered surfers to convene in rising mountain winds. The ensembles of wired brain activity can also be continuously shaped by new experience and consolidated during sleep. The nucleus accumbens is strongly interconnected with the striatum, and the striatum is connected to the thalamus through the globus pallidus. And, while sounding like a familiar anatomical rhyme, this sort of forebrain-striatum-thalamus-forebrain loop is also involved with song learning. An expectation—in this case the expected result of a social behavior like play—can be reconsidered in the forebrain by comparing it to the actual sensations of play. The playing crow's brain compares what it expects to what it experiences and adjusts the bird's behavior to the current situation. This also happens at night, without physical action, as the sleeping crow dreams about play. As the playful ravens dropped their bark surfboards for the night and glided to the valley roost, they might even have relived the thrill of riding the big wind above the scenic overlook and refined their moves for the next windy day.

While we are far from understanding all the complexities of how the brain motivates, rewards, coordinates, and shapes a vigorous behavior like play, many parts of the brain and its chemically regulated neural circuits are clearly involved. This is the key in understanding perhaps the fundamental reason that birds, and probably we humans, play. Play is a cerebral workout. Players must coordinate emotional and physical responses to their living and material environment. By doing so, they craft critical neural connections among essential parts of the brain involved in memory, emotion, sensory integration, and evaluation of the costs and benefits of actions. Developing neural play circuits provides an animal with the brain resources that can later nuance actions to the behavior of others in various social settings. These mental abilities are useful in guiding a crow or raven through assertive, interactive, deceptive, and appeasing interactions with its mate, offspring, neighbors, and associates encountered while feeding, commuting, and roosting. As it does for people, play for crows allows sophisticated motions to be rehearsed, while also providing the flexibility to adjust them creatively under dynamic circumstances. We—humans, corvids, lab rats, and probably all vertebrates—build better brains through play.

7

Passion, Wrath, and Grief

Crows pay particular attention to dead associates. While reminding us of a funeral, these gatherings are actually learning arenas where information on danger and social opportunity is absorbed.

In Seattle, Deborah Raymond watched as crows approached a dead bird like a phalanx of determined soldiers walking toward a fallen comrade. In turn, each crow hopped on an elevated curb to view the body and then moved on. More than a thousand miles to the north, hundreds of ravens spiraled, like debris in a tornado, above a computer store in Alaska where moments earlier two of their brethren had been electrocuted. The eerie scene lasted only a few minutes, but it was etched deeply into Rod Stephens's mind. It was as if the birds had paid their respects and then silently departed. Kay Schaffer was drawn to the window of her Dayton, Ohio, house by an early evening ruckus of crows. The birds cawed wildly as they took flight before settling in a large tree where they looked down upon a dead crow. After twenty minutes, the gathering quietly dispersed. Two weeks later, the dead crow was still untouched, but something or someone had surrounded the corpse with an outline of sticks. We authors often are told about such "funerals" and have indeed witnessed similar events.

Visiting the municipal buildings near his home one fall afternoon, Tony noticed more than a hundred crows perched silently in the sycamore trees fronting a local roadway. Many of the birds were alternately circling low over one particular part of the roadway, and a few were actually landing and wandering into the road itself. The object of interest was a dead crow, its rigid legs in the air, its wings spread limply at its sides. With a lull in the car traffic, more birds landed and approached the corpse as if to confirm its identity and fate. All the while, the birds overhead pitched out from the branches to circle low over the body. After fifteen minutes the noisy gathering of crows began to disperse, and the birds flew off in different directions. Soon the trees and parking lot were empty, and only the stiff black body remained on the road. The setting was reminiscent of a scene from a mobster movie where the Godfather's funeral brings together a somber, black-suited crowd of associates, their disputes temporarily set aside as each views the body to pay his respects.

Crows and ravens routinely gather around the dead of their own species. Rarely do they touch the body, in striking contrast to their reaction to the dead bodies of other species, upon which they quickly feed. Of course, they recognize their own kind, and in most cases they probably recognize the individual that has fallen. But it's not clear whether they are paying emotional respects or simply using the moment for individual gain. In the mobster analogy, even as they pay their respects, members of the gathering are eyeing one another to determine how this death might advance their interests. Perhaps these very social birds are also working out a shift in their hierarchy. The momentary void will soon be filled, and navigating this change could provide benefits in the acquisition of mate and territory. The assembled birds may be assessing how they fit into this new social hierarchy as well as investigating the cause and circumstances of death and how they might avoid a similar fate.

Learning from fateful situations is common among animals. It enables crows to learn about particularly dangerous people, an ability we relied on when we trapped Hitchcock, the windshield-wiper bandit, and his mate in the North Cascades. As crows or ravens gather around a dead mate, offspring, neighbor, or stranger, they are acquiring knowledge about danger and opportunity. This learning is most likely facilitated by the sensitivity of their brains' amygdala to the release of stress hormones.

In mammals, including humans, the amygdala is at the center of evaluating sensory information of social and emotional significance. Just as the amygdala is important to sculpting what one learns through play, it is critical to the attention, perception, reasoning, and remembering that occurs during sad or fearful events. The amygdala tunes our attention to emotional scenes through its reciprocal connections with the hippocampus and visual areas of our brains. Even before we are consciously aware of an event, the neurons in our amygdala have been activated by visual sensations, which trigger reflexive responses that avoid danger and focus our attention on the scene. As our forebrain considers danger, the visual cortex increases our perception of details, and the hippocampus provides historical context.

When quail encounter a predator, they often freeze to avoid detection, and this behavior appears to be regulated by the action of the amygdala and its connections to the social brain network. So we suspect a similar response by a crow. A bird fearful or saddened by the sight of a familiar, dead individual would attend closely to the scene, and its heightened emotion would influence its perception and attention to detail. This focus enhances learning, about danger and about opportunity.

As a crow views the deceased, the potential danger of the setting or the sadness conveyed by a loss of a companion would stimulate its adrenal glands to send corticosterone and epinephrine into the blood stream. These stress hormones dampen play behavior and enable linkages to be made between sensory and amygdala neurons. This helps them forge new associations among regions in the social brain network that represent learning about potentially dangerous situations. Stress hormones quickly pass from the blood into the brain and synergistically aid memory formation by binding to receptors on neurons in the amygdala. In a stressed crow, epinephrine binds to neurons in the vagus nerve stimulating the release of norepinephrine into the amygdala. The more distressing the stimulus, the more norepinephrine is released and bound to neurons. The same is true for corticosterone. Simultaneously, these two hormones increase the intensity and strength of electrical signals passing from the amygdala to other parts of the social brain network, the ease with which the next dangerous sight will trigger a similar response and the likelihood that those neural connections will be stored as memories. Just as a playful crow learns and remembers, a mournful or frightened crow also learns important lessons about its safety.

While the cellular basis of memory formation—the differential strengthening and proliferation of synapses—is the same whether learning takes place while an animal is playing or under stress, it is the binding of various chemicals to neurons that determines what synapses will be coalesced into pleasurable versus distressing memories. Dopamine is released from the brain stem in response to surprising stimulation, whether pleasurable or depressing, and increases the formation of memories involving neurons from many brain regions. This process is likely at work in a crow that is learning while at play or in a crow that is learning while in distress.

The influence of opioids, on the other hand, may help determine which neurons from the amygdala are consolidated into playful versus fearful recollections. The release of opioids during play reduces the binding of norepinephrine to neurons in the amygdala, which should limit the formation of strong connections in the nucleus accumbens between the amygdala and other brain regions. In contrast, during stressful situations, opioid release is suppressed, and the influence of norepinephrine on neurons in the amygdala is less constrained so that more, and possibly different, connections can be made between the amygdala and other brain regions. We expect that the neural ensemble—the memory—that forms as a crow views a dead companion would include many strongly connected wires to a very active amygdala, whereas a memory that formed as two crows played would be less, or at least differently, wired to a less active amygdala.

We are convinced that crows and ravens gather around their dead because it is important to their own survival that they learn the causes and consequences of another crow's death. We also suspect that mates and relatives mourn their loss. In humans, mourning serves to celebrate a former life and to reduce sadness, yearning for reunion, and intrusive thoughts about the deceased. The same could apply to crows. Human grief appears to be regulated by the neural connections between the amygdala and the integrative centers of the prefrontal cortex. Reminders of the departed stimulate the amygdala to recall memories of the loss, but as time goes on, grief-inspired introspection strengthens the feedback from our prefrontal cortex to our amygdala, reducing its response to formerly powerful reminders. Crow memories of death are likely formed in much the same way as ours are formed, so it is certainly possible that, as crows revisit their memories to adapt to possible dangers or new social opportunities, they are also saddened.

If crows mourn, this emotional state is rather short-lived, judging from our observations of their close relatives, the jays. In pinyon jays, as in most corvids, lifelong monogamous pair bonds between mates are the central organizing relationship in their society. Yet, when a partner is lost, another is quickly found. In 1983 John was watching a pair of jays in their eleventh year of monogamy tend a recently laid clutch of four eggs. The morning after a heavy spring snow he noticed the nest had been abandoned. Blaming the storm, he expected to soon find a new nest built by this long-standing pair, but the thirteen-year-old female was dead. Her mate wasted no time in joining another recently widowed female. Within days of his mate's death he was tending to his new mate, even helping raise her existing brood of five chicks.

The fast pace of a bird's life may not be compatible with extended grieving. A quick acknowledgment of death and a rapid social adjustment followed by the continuation of life may be all that is possible, and such action would certainly favor the individual intent on passing its genes to another generation.

A world away, in the wilderness of Yellowstone National Park, bison grieve. On a crisp March afternoon on a snow-dotted sagebrush flat, John's students and colleagues were engaged in a survey of a recent wolf-kill site. The victim, two weeks dead, was an old female bison. Her bones had been scattered among the level terrain, picked clean by coyotes, foxes, and ravens. On her final day, the bison was pursued by a pack of wolves, and seeking shelter, she wedged herself into a nearby boulder, split in half by eons of freezing and thawing. Her blood still stained the inside wall of each half boulder, revealing to us that the cleft rock offered only partial protection. The wolves attacked from the front and the back. Gruesome, but no doubt better than suffering through the winter; previous observations by the biologists in Yellowstone revealed that this matriarch had suffered from a detached placenta.

As we collected our information, a sudden thundering of hooves pushed us from the skeleton. A bison herd roared down at us and stopped to view the remains. As we watched, each of the three dozen animals walked up to the female's bones and smelled them. They sniffed the remains and the soiled snow and dirt. In all, they remained at the site, keeping us from it, for nearly an hour. Then they walked through the cleft boulder where the killing had occurred and filed out of our view over a low shoulder of tranquil turf. All of us felt privileged at the experience, guilty at our indifferent attitudes toward another species, and convinced of the sacrosanct nature of the procession we had witnessed. These animals were still sensitive to a past event, behaving in ways reported by scientists who study elephants, monkeys, and apes. In all our years watching birds, including watching many die, never have we seen such a demonstration of prolonged concern.

Differences in grieving behavior between jays and bison might reflect differences in the neural chemistry or wiring in bird and mammal brains. While we do not know how the brains of monogamous birds reflect their social bonds, we do know quite a bit about how this occurs in mammals. In particular, we know the neurobiology of pair bonding in voles—those mouselike, sausage-shaped mammals with short tails and diminutive ears. In fields, these rodents burrow under snow and grass, leaving crisscrossing runways that are often revealed when the weather warms and the snow cover melts. The prairie vole is among the most monogamous of its species. Again, it is a brew of chemicals including dopamine and two hormones, vasopressin and oxytocin, that interact in the nucleus accumbens, septum, pallidum, and prefrontal cortex of a vole's brain to guide its monogamous nature (the binding of chemicals to neurons is illustrated in the Appendix, pages 230–231, as are the brain regions important to social behavior).

The neurobiology of monogamy is well understood in the prairie vole.

To better understand crow behavior, let's look at how chemistry affects the behavior of voles. Dopamine appears to be important to the formation of pair bonds and is released into the nucleus accumbens in association with sex, just as it is released in response to other rewarding activities like playing or eating. When dopamine binds to one type of receptor (called D2) in the nucleus accumbens of a prairie vole, it motivates pair formation in males and females. Yet when dopamine binds to a different receptor, (D1), it specifically reduces preferences for other voles by mated males. In this way binding of dopamine to D1 receptors in male vole brains prevents males from seeking additional bonds beyond the one established with a female sexual partner. This effect is long lasting; after a male breeds with a female, the density of D1 receptors in his nucleus accumbens actually increases. His neurons are modified to bind more dopamine to D1 receptors after sex, thereby precluding the formation of additional pair bonds and reinforcing truly lifelong monogamy.

Hormones, opioids, and dopamine conspire in the brain of a vole to increase desire for, and recognition of, a single mate. The hormone oxytocin generally increases bonding in mammals. In humans, it promotes the mother-infant bond when it is released during breast feeding. In female prairie voles, dopamine and oxytocin reinforce fidelity by binding to neurons in the nucleus accumbens, which reduces desire for additional partners and reinforces attraction to her mate. A male vole is controlled in the same way by a brew of dopamine and the hormone vasopressin. Hormones also increase a vole's ability to recognize its mate. When oxytocin is prevented from binding to neurons in the amygdala and vasopressin is prevented from binding to neurons in the septum, a vole no longer recognizes its mate. These hormonally responsive neurons come under the influence of opioid rewards in the nucleus accumbens. Male voles learn to associate the opioid reward of sex with the scent of a particular female, so that after sex, simply smelling the female reinforces his fidelity.

This helps us understand the apparent differences between grief in jays and bison, because birds also produce hormones that are molecularly similar and functionally equivalent to those found in voles. They are called vasotocin and mesotocin. Mesotocin promotes social bonding mostly in females, and vasotocin does likewise mostly in male birds. So, like mammals, birds have sex hormones that interact with dopamine to affect social bonds. In fact normally social zebra finches become distinctly solitary if mesotocin binding in the septum is blocked. We suspect that the binding of dopamine and sex hormones in the nucleus accumbens, septum, and amygdala are also key determinants of pair-bond formation in birds as they are in mammals. But the fact that monogamous birds, unlike prairie voles, readily pair with a new partner after a mate is lost suggests that there may be subtle neural differences in how bird and mammal social bonds are reinforced. Perhaps bird social bonds do not involve changes in D1 dopamine receptor density in the nucleus accumbens, for example. Reduced dopamine binding after the loss of a partner might quickly motivate search for a new mate. Or, perhaps using sight and sound to choose a mate, which is the norm in birds, is less closely linked to the reward of sex than is choice based on scent, as practiced by most mammals. Such differences in the neural constructs that represent bird and mammal social bonds might explain the differences we've seen in pair bonding and, by extension, the grieving we observe when important social bonds are broken by death.

Two crows were observed supporting an injured associate.

When confronted by death, crows are restrained, thoughtful, and, on occasion, passionate. In a typical suburban neighborhood, just across Lake Washington from Seattle, a retired medical doctor watched as three crows hobbled across a lawn. The center corvid's wing drooped lame, likely broken, as the other two birds pressed close, one on each side. The two outer birds appeared to be supporting the hurt bird with their wings, just as two GIs might carry a wounded patriot from the battlefield. The trio hopped past the stunned doctor and disappeared into a nearby forest.

A crowd of golfers, including Chuck Wischman, witnessed similar heroics on Seattle's Jefferson Park Golf Course. As they prepared to tee off, the men heard a loud "pop" and turned to see a crow coldcocked by an errant approach shot to a nearby green. The players thought the bird was dead as it lay face down with its head slightly uphill. Immediately another crow began calling loudly, pecking around the fallen bird, and pulling on its wings, which flipped the bird over and oriented its head downhill. Five other crows were attracted to the calling, and three began pecking and pulling on the apparently dead bird, trying to lift it up by the wings. After a few minutes, Chuck and his partners were convinced the crow would not survive and turned away from the heart-wrenching scene to resume their game. Other golfers stayed and two holes later delivered the stunning news that the crow revived, fluttered briefly, and then flew off.

Often, when a crow is injured, the response by the flock swings wildly from compassion to rage. A zoo researcher observed the typical reaction to an injured Seattle crow. A recent fledgling was grounded but well protected by a growing flock. As Carol Strickland and her dog approached, the crows grew aggressive and protective. Their actions kept human and canine at bay. Minutes later the protective crows suddenly switched demeanor and killed the young one. It seemed like a mercy killing, swift and accurate, ending a life certain to suffer.

Reports of crows maiming or murdering a flock member have been sent to us from Austria, Seattle, Vermont, Nova Scotia, British Columbia, Massachusetts, and New Jersey. Barbara Brozyna's observation is one of the most complete. During a cool March morning of 2004 in Toms River, New Jersey, a small crowd of noisy crows captured Barbara's attention. Across the street from her, the boisterous gathering directed their clamor toward a single crow that stood silently apart from them. Five or six of the crowd moved away, leaving three to continue cawing and facing the loner. Suddenly one of the three attacked, pummeling the lone bird with powerful pecks. The assault continued for several minutes as single birds one after another stabbed at and then retreated from the victim. After about a dozen rounds, the outcast was bloody, lying on its side, legs reflexively twitching. The trio flew to a nearby tree and silently looked down on their near-dead victim. In one last effort the injured bird struggled to a protective cedar, but the assailants swooped in pursuit and finally killed the wounded bird.

What possesses a crow to kill? Might subtle changes in brain chemistry underlie such drastically different responses to flock members? We've all experienced how quickly our emotions can change, depending on the environment or signals we receive from social companions. The same is possible for a crow, who has similar brain structures and a similar neurochemistry of emotions. While defending territory, guarding a mate, or chasing a predator from near a nest, sex hormones like testosterone or stress hormones like epinephrine bind to neurons in the birds' preoptic area and hypothalamus, causing a cascade of electrical and chemical signals through the brain that typically result in aggression or sexual behavior. These actions would push a crow to attack a competitor or a predator. Such motivations probably preceded observations of one crow killing another. A parent's aggression toward a nest predator might carry over to another subject, even if it was a wounded fledgling or grounded nestling. A hormonally enraged father might stun the intruder, leaving it vulnerable to future attack.

Most of the time aggression stops when the predator leaves or the intruder submits, but this critically depends on opioids in the aggressor's brain. If opioids bind to the neurons firing in response to testosterone, for example, the rate of firing drops. Opioids cool an aggressive bird's jets, just as they deaden our pain, but they are released in response to attaining a goal—evicting a predator or an intruder, for instance. If the goal is not attained—perhaps the predator remains and eats the nestlings that did not escape from the nest, or a stunned intruder is unable to leave or signal submission—opioids are not released, and aggression spirals out of control. We guess that events prior to the killings or rescues that are observed are critically important to the emotional state and resultant behavior of crows, but unfortunately these preceding events are unrecorded. It seems certain that a helpful crow's brain is stimulated by chemicals like mesotocin and endorphin, while a murderous crow's brain is not. But why this varies in nature remains a mystery.

Crows and ravens do not require powerful physiques to apply their wrath. Without the penetrating, crushing, or strafing talons or hooked beak that hawks, falcons, or eagles have, some corvids use their brains to overcome their physical limitations. As two young white-necked ravens challenged each other for a bump of high ground, one bird picked up a stick. As the other charged, the tool user hurled the stick to defend his position.

In another story of a corvid wielding a weapon, Madeleine Kornfield was walking across the University of California, Santa Cruz, campus on an early spring afternoon, when she noticed two ravens confronting an owl perched in a live oak tree. As one raven bounced on a flexible branch above the owl, it caused the oaken weapon to repeatedly crash onto the drowsy predator. Then, the second raven came in from below, shielded from the bouncing branch. This bird brandished a small twig and began poking at the owl. Students gathered to watch the show of cooperative brain versus brawn, which after about five minutes was won by the smart ravens. The owl flew off to a less accessible retreat.

Ravens use sticks and branches to taunt a great-horned owl.

When two corvid species resort to using weapons against each other, the results are less predictable. Noted ornithologist Russell Balda was enjoying a spring morning in Flagstaff, Arizona, watching a lone American crow breakfast at his bird feeder when a pair of Steller's jays arrived. One of the jays began scolding the crow, rushing at it and retreating as the crow turned to face and lunge at the much smaller jay. The jay swooped at the crow twice from a perch high on the Baldas' roof, but the crow held its ground and continued to steadily scoop up seeds usually taken by the jays. After the second swoop, the jay flew into a nearby mountain mahogany bush and twisted off a four-inch-long pointed stick. With the sharp end facing forward, the jay held the stick in its beak and lunged toward the crow. The joust barely missed the crow, who lunged back at the jay, causing the weapon to fall onto the feeder. The crow recovered the stick and, as the jay had done, gripped the dull end, aimed the sharp end toward the jay, and lunged. That was effective. The jays flew off, and the crow followed in hot pursuit, stick in beak.

A sequence of behaviors observed at an Arizona bird feeder. A Steller's jay used a stick as a weapon to evict a crow, but the substantially larger crow turned the tables, took the stick, and chased the jay.

Crow rage is impressive, but rare. A fight between rivals is much more likely to end in consolation and reconciliation than murder. In captive groups of ravens, rooks, and Eurasian jays, when two members of a flock fight, the victim is consoled by an onlooking, important social partner. When close associates, such as relatives, fight, they quickly reconcile their differences and repair their bond. Consolation and reconciliation often involves nurturing interactions, such as mutual mouthing and grooming that reduce stress and calm participants, reducing aggression and restoring peace to the flock. The victim's brain is likely soothed by endorphins, not unlike the chemical response of a boxer consoled after a tough round by his trainer's encouraging words.

Corvids may even console people. Esther Woolfson was weeping, when her pet magpie, Spike, responded to Woolfson's grief with a submissive posture and flight to her knee to nuzzle her and softly call. The social milieu may greatly influence individual behaviors, something difficult to assess in nature where large freewheeling flocks include relatives, mates, allies, enemies, and strangers. An interaction that captures our attention is a product of past associations and present experiences, all of which tune a crow's brain to respond effectively to its changing and challenging circumstances.

The most intriguing stories that we hear about corvids challenge our rational scientific minds. We think they reflect human culture and the response of corvids to human emotion, but to those who tell about such events, the crows' metaphysical significance or spiritual power is real. But our travels into the minds and emotions of crows have also alerted us to the fact that our own emotions, experiences, and expectations influence how we perceive and remember our world. So let's explore a bit beyond scientific evaluation of crow behavior to consider some stories of crows with extraordinary influences.

In India some people believe that crows are their ancestors. They revere these birds. In an ancient custom after a person dies, his or her survivors divide the spirit of the departed into a number of rice balls. It is believed that the ancestors come as crows to eat the rice balls and liberate the soul of the departed. If, however, the deceased had unfulfilled desires, the crows will not eat the rice. A mother in

Kerala, India, had hoped to see her favorite son one last time, but her spouse had disowned the son and would not tell him of his mother's death. The crows would not eat the rice offered to them. The son did hear of his mother's demise and returned in time to offer a late afternoon rice ball to the crows, which they ate, taking his offering first and then those they had earlier refused.

Nearby in a spiritual ashram in Amritapuri, the faithful gather each evening for two hours of devotional singing. Just as the singing starts, shortly after crows normally have gone to roost for the night, a lone crow regularly flies into the ashram and perches on one of the large hanging chandeliers. At the end of each song the bird changes position, moving among the three lit chandeliers. When the music ends, the crow departs and returns only the following night.

Many Native Americans and others who are of a pantheistic inclination pay close attention to nature, believing that spirits take the shape of animals in order to convey important messages. There is a practical side to this spiritualism. A few years ago we visited with Tlingit artisans Demsey Bob and Norman Jackson in Ketchikan, Alaska, who told us that their elders would calm bears by telling them they were only here to pick berries and that they would improve their hunt by asking ravens where the moose are.

Among native people, the presence of a raven or crow is never taken for granted. Kent Bush, a former regional curator for the Pacific West Region of the National Park Service told us of his Navajo colleague, maintenance chief Tommy Dempsey, who reacted to seeing a raven one morning as they drove toward Canyon de Chelly in northern Arizona. Dempsey noted a raven following the road and said: "There's brother raven doin' his morning road check—don't scare him or we'll have trouble the rest of the day." Leaving the raven undisturbed, they enjoyed a trouble-free day.

Cathleen Handlin, a cultural anthropologist with the U.S. government, was impressed by the genuine respect for crows shown by a Native American dancer demonstrating his skills at the California amusement park Knott's Berry Farm. The dancer was showing tourists how he was making a feather headdress for the crow dance. The regalia included many turkey feathers, dyed black to simulate those of a crow. Suddenly a wild crow started to call as if it spied a dead crow. The harsh calls attracted a mob that scolded the black feathers, and the culpable humans. The performer was mortified at what he had done to his crow brothers. He announced to the crowd that the crows were hurt by the possibility of the carnage, he was to blame, and because of this he was stopping his presentation for the day and would not include black feathers in future shows. He gathered up his handicrafts and beat a hasty retreat.

Many believe that the voices of corvids are sources of inspiration. Jim Schumacher, an eminent marine biologist, was walking his dog one afternoon when two crows attracted his attention. Spiritually inclined, he listened, wondering what information the crows were conveying to him. He sensed they were asking him to change his name. Previously Jim had taken the name of Little Black Eagle, which is Cheyenne for "crow." After listening, he became "Two Crow." Dr. Jim Schumacher authored his last scientific publication in 2005 under the name of Two Crow. A practicing shaman in southeastern Ohio, Crow Swimsaway, has a tattoo of crows on his arm. He holds a PhD in social anthropology from the London School of Economics and also received his spiritual and legal name from a crow.

Most of the people who experience unusual corvid encounters sense a connection, but not as strongly as Two Crow or Crow Swimsaway. For example, Dana Casey was just enjoying an afternoon view over the Pacific Ocean from Kalaloch Beach in northwestern Washington when she noticed a flight of ravens drifting on the sea breeze. Quite suddenly and inexplicably, she had an out-of-body experience soaring with the birds. Never before or after has this sort of thing happened to her.

Mary Coggins, an end-of-life caregiver from Port Townsend, Washington, often notices crow corpses and is convinced that crows go out of their way to point her toward their fallen. As she visited one family whose father was dying, she noticed a crow prancing and bobbing just outside the French doors. The bird surprised the family, as they had not seen a crow in the area for years. Mary told them that crows seem to seek her out. A son-in-law of the dying father was just heading out the door, but returned, ashen at having seen a dead crow on the front lawn. To Mary, crows affirm her chosen work and spiritual path.

The connection between crows and death pervade many human cultures. We often hear that the presence of a crow or the behavior of a raven foretold a human casualty. Tom Trent, a former assistant ravenmaster at the Tower of London reported to our friend Boria Sax that ravens sensed the death of the resident chaplain Norman Hood in 1990. After Hood died in his chamber at the Tower, the ravens soon gathered on the Tower green, near the chapel. They called out and then went silent.

Suzanne Wyman of Vancouver Island, British Columbia, communes with the eagles and ravens that surround her coastal home. She told us that crows signaled her husband's passing. In May 2001, as Suzanne walked along Cowichan Bay, she noticed three crows sitting in a tree. While this alone was not unusual, these birds were aligned in such a way that they formed a triangle and were intently staring at her. Returning home she sketched the image in her diary. The next day as she went to Victoria General Hospital, an hour and a half away, to visit her terminally ill spouse, she again saw three crows outside the hospice door perched in a tree. She had not seen crows there on any previous visits, and they alarmed her because they were sitting in a triangle configuration, just as the ones on Cowichan Bay had done. Inside, she learned that her husband, Rob, had died.

Perhaps there is no rational explanation for stories such as these, and there doesn't have to be one. We accept each story as reliable, having interviewed each person, but we are compelled to suggest that human culture and the human brain have conspired to create their views.

Corvids pervade human culture. Our art, language, religion, and ways of knowing have been shaped and continue to be shaped by our experience with crows, ravens, and other jays. They have come to symbolize death, to speak for powerful spirits, to guide, inspire, inform, and engender wonder. These are powerful attributes, and they reside in our brains, affecting our perception.

Benjamin Libet, a respected neuroscientist from the University of California, San Francisco, spent his lifetime collaborating with surgeons to understand better how our unconscious sensing of the environment becomes our conscious experience. By simultaneously timing patients' neural reaction to a stimulus and their conscious awareness of the stimulus, Libet discovered a consistent mismatch. Patients did not report the first neural response their brains made to a stimulus—a pinch, a loud noise, or a child running in front of a car. Rather, after only a half second of persistent firing by neurons in response to a stimulus did patients report being aware of the stimulus.

Our senses stimulate our brains a half second before we are aware of the stimulation. This delay is our subconscious. Part of the subconscious, but a very small part, is the time needed to relay information from the sensory organ to the brain or spinal cord and back to the muscle for physical response. This is our reaction time. Most of the delay is due to the requirement of persistent stimulation—a train of action potentials in a neural circuit—in our forebrain before a mental representation of the event emerges. As weird as it sounds, we are actually living in the past. Our conscious experience is a consequence of persistent, not immediate, neural activity. Subliminal advertising works because the short stimulation influences our subconscious, not our conscious; we get the message, and our bodies might react, but we are unaware.

Think about a child dashing in front of your car. Your retinas sense movement and send action potentials toward your brain. Reflexively, your brain commands action, and you stomp on the brake pedal. That takes about 0.15 seconds. It took another 0.35 seconds before you became consciously aware of what you did.

The brain is slow to incorporate stimulation into conscious awareness, but it is active during this delay. The cerebrum is considering and sculpting sensation into awareness. Incoming neural signals are affected, and synapses are strengthened or weakened by hormones and neurotransmitters. The amygdala sculpts perception with emotion. The hippocampus and other memory stores sculpt perception with history and geography. This adjustment of stimulation allows us to suppress unpleasant sensory information, focus on novel information, and alter the conscious content of each event we observe. While we mentally live in the past, our minds compensate for the delay so that we think we are immediately aware. But the delay is real, and it allows our individual experiences to affect the translation of our perception into our reality. To quote Professor Libet: "Each person has his or her own conscious reality."

The brain of a crow also integrates and reconsiders memories to shape its mental construction of the outside world. These expectations temper how the bird responds to new information. It may be crucial to the crow's adaptability. The brain of a person is profoundly more capable of reconsidering and sculpting what it senses into what it knows. Our processing allows us to navigate a high-speed, multifaceted cultural and physical world. It also uniquely tints each of our worlds, because our reality, in part, reflects our memory. So we suggest that Dana's reality did include soaring with ravens and Jim did understand a pair of crows. Further, Mary's occupational immersion with terminal care influenced her perception of crow death, and Suzanne's cultural familiarity with the crow as a harbinger of death adjusted her perception of the birds outside her husband's hospice door. All of these observers shared a history of profound interactions with crows. Their histories subconsciously changed what their eyes saw into what their minds experienced. Other people, without recent experiences or cultural connections to corvids could have stood next to Jim, Dana, Mary, or Suzanne and never have heard a crow speak, flown with a raven, noticed a crow corpse, or seen a trio of crows beside the hospital door.

8

Risk Taking

Sometimes mobbing a dangerous predator ends in death.

BALD EAGLES have a stately elegance and restrained power. Here in Washington they can be found perched in rough cottonwoods, scavenging spent salmon, or flying low over our suburban neighbors. These impressive birds have made a strong population recovery following the banning of DDT in the United States and a comprehensive effort to protect their feeding and breeding habitat. Today they are often sighted within many of our country's largest cities and along busy freeways.

While an inspiration to us, to crows, eagles are the enemy. Efficient scavengers, eagles are also skilled predators. Their nearly seven-foot wingspan buoys a ten-pound body, while the tip of their penetrating beak can shred a frozen elk. Their toes are equipped with talons up to an inch long. Mates team up to hunt waterfowl and fish, and in the spring their diet includes nestling crows and the occasional adult.

In comparison, a crow is weak—a David to the eagle's Goliath. Their needle claws scratch and hold but do not kill. Their beaks are versatile but unable to subdue much of anything larger than a rat. Yet where crows and eagles coexist, one routinely sees a murder of crows dive-bombing a lone eagle. Cawing loudly and spiraling up above their adversary, they begin to line up like aircrafts staging for a bombing run and one by one dive relentlessly at the foe. The eagle labors along, seemingly indifferent to the enraged birds around it. Sometimes the hunter is simply driven away, but crows do occasionally hit and even kill eagles when the severe impact of a one-pound black missile collides directly with the larger bird's head. More often a crow dips too near and is snatched reflexively in midair, soon to be eaten.

Risking one's life to harry an eagle, owl, or hawk seems a required duty for many corvids. This behavior is a calculated risk, one favored by natural selection simply because the gains outweigh the costs. Harassed predators are usually distracted from hunting and made obvious to their prey. They're soon off to find a less hostile environment where the element of surprise will be in their favor.

Chasing a predator out of a nesting or roosting area or reducing its chance of success benefits a mobber. Members of species that mob live longer than those of species that do not. Mobbing may do even more; as a social exercise it provides a forum within which individuals can display their flight skills and aggressive tendencies. Not unlike the human recognition for a demonstration of valor, these attributes may win a daring crow a mate or a higher rank in its avian hierarchy. Naive crows can learn about individual predators by observing a mobbing.

While some risky behavior is easy to explain when we balance all the costs and benefits, other activities seem downright foolish. Glenn Rimbey was riding with his son, heading south on US Interstate 5, just north of Sacramento, California, in early May 2006. At highway speed, they spotted a raven dead ahead, in the middle of their lane, engrossed in a freshly killed animal. As they passed directly over the bird, the Rimbeys cringed, certain they had added a bit of corvid body to the undercarriage of their vehicle and a bird body to the roadkill on the asphalt. Looking back to confirm his worst fears, Glenn was amazed to see the raven rise up from a crouch and resume his meal. As another car approached, the bird again held its position, squatting to the pavement, lowering its head, and allowing the speeding car to pass safely overhead. The smallest mistake by the bird or the driver would certainly extinguish this risky foraging strategy. Yet this raven seemed to have beaten the odds and learned a new way not only to procure a meal but also to continue its feeding relatively undisturbed. For the time being, the access to food that others dare not come near makes the risk worth taking.

Risks can pay off. This raven learned to hunker down as cars sped overhead, allowing the bird access to roadkill others dared not touch.

The bounty of food found on the road seems to push corvids to scavenge near certain death. Crows routinely feed on roadways but usually just beyond the heavy traffic lane. Robert Mooney, a longtime observer of crows along Interstate 77 in Ohio, suggests that crows have habituated to the buzz of traffic. In the 1950s, Robert noticed crows flying away from the road shoulders, but by the 1960s they seemed only to turn their heads to protect their eyes from the gush of wind and debris created by the cars that whizzed past. Certainly they realize the danger of moving cars and quickly acclimate. To most corvids this opens up the food bonanzas of the highway medians and edges. To some individuals, like the raven Glenn observed, it may also open up the portion of the driving lanes straddled by crushing tires.

In Japan, some crows even learn to associate traffic signals with safety and danger. Jungle crows stride across pedestrian crosswalks when red lights stop traffic but perch above roadways and watch when the light turns green. Some of these birds take advantage of the stop-and-go of traffic to place nuts too large to crush with their beaks on the asphalt, but most seem to search the road for its natural bounty of roadkill, squashed bugs, and food bits dropped by people.

The benefits of working the road were much less obvious when Matt Betts, an ornithologist at Oregon State University, encountered a crow on a road in Corvallis, Oregon. Professor Betts was bicycling to work in the autumn of 2008 when he noticed a crow lying in the middle of Twenty-Ninth Street. As any good biologist would do, Matt went over to inspect a possible specimen. But the crow was not dead. The day was bright, the tarmac dry, and no food was evident to either confuse or lure a crow. Assuming the bird was injured, Matt picked it up and moved it to the safety of the road's edge and noticed a large tumor around the eye and bill, possibly crow pox. When he put the crow down, it seemed strong and far from death; it perched upright briefly, flew, and then circled around in a coordinated flight before returning back to lie in the middle of the road. A second rescue ended in another return flight to the road. And so did a third, after which the professor concluded the bird preferred the road, and so he rode on to his morning appointments.

What might account for the maladaptive, risky, if not suicidal, tendency of this bird? We suspect the tumor was somehow involved. Perhaps the disease affected the bird's ability to correctly sense its environment. While its brain issued proper commands to flight muscles, it may have done so without accurate or complete visual imagery. Alternatively, a previous injury may have damaged the crow's amygdala, which has been shown to increase fearlessness in birds. Or a previous bruise of the spinal cord may have disrupted the rhythmic activity of neurons that connect the thalamus and the forebrain. The firing activity of these neurons regulates the alternation between states of sleep and vigilance, which appeared disrupted in this crow. There was no apparent benefit to lying in the road, so we doubt it was a purposeful choice. Instances of apparent suicide among birds have not been reported, so the evidence leads us to conclude the crow's brain was not fully functioning.

Must all risky behavior be purposeful? For the most part, yes. In nature an animal's actions contribute to its reproduction and survival. Risky behavior may persist in a population if it has benefits. Most overly risky individuals will likely soon die or fail to reproduce. In European forests, for example, when beechnuts are rare, aggressive female great tits (birds related to American chickadees) outcompete shy females for limited foods and survive better over the winter. The tables are turned on the bold tits in rich years, however, as the shy females, who fight less, survive better.

Over time, natural selection removes inappropriate risks from a species' behavioral repertoire, or a species makes adjustments. When Amtrak initiated high-speed train service in New Hampshire, for example, a crow who was feeding on spilled grain in the tracks was killed. Crows apparently routinely fed in front of slower freight trains, waiting until the last minute to fly out of harm's way. This risky behavior was quickly attuned to the new speed of commuter trains. Crows continue to scavenge from the tracks, but now they fly sooner; perhaps they learned from earlier birds' miscalculations or those crows with a more precise sense of timing were spared.

Behaviors that seem risky may actually happen in relatively safe situations. For corvids, safety often comes in the form of human shelter. Living in our houses, yards, or cities, corvids have superabundant food, relatively benign climate, and few predators. Under these conditions, risks taken may go unpunished. Without punishment and under the influence of the brain's pleasurable rewards, animals can form risky alliances, like crows that adopt kittens and ravens that enjoy a good romp with a dog.

Feral and free-ranging pet cats threaten birds wherever humans settle. Domestication has not taken the predator out of felines. Worldwide, cats are implicated in the extinction of thirty-three species of birds. An invasive species when let outdoors, cats are estimated to kill nearly 500 million birds each year in the United States alone, as well as many native small mammals, amphibians, and reptiles. Crows are far from immune to the wrath of cats. On Hawaii and the Northern Mariana Island of Rota, for example, the combined remaining number of two spectacular species of native crows totals fewer than three hundred individuals, in part because of cat predation and in part because of the deadly diseases, like toxoplasmosis, that cats spread.

Despite the obvious tension between cats and birds, some crows bond with cats. Porter Evans's deaf and partially blind tortoiseshell kitten was alone in the backyard when a hawk suddenly bore down, senses locked on to the cat, talons extended, and only seconds away from a furry meal. Then something extraordinary happened. Three crows, regularly fed by Porter's mother, flew up and at the hawk, deflecting its predatory trajectory and sparing the kitten with which they had a peaceful relationship. The Evanses always welcomed crows that regularly ate from a custom-built feeder. Ms. Evans even had rescued and rehabilitated one fledgling. Despite the kindness attributed to crows by the Evanses, the crows' response to the hawk was likely selfish; they probably never even noticed the kitten, but by intercepting the hawk, they were increasing the safety of their territory.

Another crow-and-kitten story, televised widely and viewed over seven million times on the internet, is more perplexing. Mr. and Mrs. Collito from Massachusetts filmed a stray, mostly black kitten and a young crow interacting in 1999. The crow, molting and with a pink mouth lining indicative of a fledgling in its first autumn of life, walks side by side with the kitten. The crow appears to feed the cat worms and other natural items, pulls at the cat's tail, wrestles with the playful cat, and is pounced upon by it and remains an ever-present companion. The Collitos are sure the abandoned kitten would not have survived without the friendship, sustenance, and protection the crow provided. After the Collitos adopted the kitten, they began keeping it inside at night. The crow would caw at the window each morning until the Collitos let the cat out to play.

Dogs are also predators, though less of a threat to birds than cats are. In our yards, where crows often thrive, they must learn the ways of dogs. In our field research we occasionally place radio transmitters on crows so that we can follow their every move and observe how they interact with predators, such as dogs. As John homed in on one recent fledgling, he was able to follow the young crow's daily challenges, observing how this crow and its siblings learned from their parents about bounty and risk, which were everywhere. Cars plied the roads bordering the suburban crow territory. A Cooper's hawk nested only a football field away. Coyotes kept the cats scarce, but as wild canines, they, as well as native bobcats, would gladly eat the young crow. Our radio crow provided a lesson to her brothers but learned little. She was only four days out of the nest when a German shepherd, whose territory included her nest, caught and killed her.

Young crows learn about risks and rewards by observing their parents and siblings. Young hooded crows watch others mob a barn owl.

In contrast to the typical collision between wild crows and domestic dogs, there are instances of surprising partnerships that appear to be rooted in friendship. Jeanette Griver has written extensively about her Shetland sheepdog's escapades with a crow in California through a series of novels. The dog and crow, like the crow and kitten from Massachusetts, were inseparable. They played together and especially seemed to understand each other's intentions. When the crow wanted to play, it would approach Griver's door, attract the dog's attention, and then caw until the dog was released. Mutual understanding of intention is further supported by a photograph by John Hoeschele of a young crow and a wiry, black Labrador mix. The dog is shown crouching forward, haunches high and tail wagging in typical play bow posture, inviting the crow to join the game.

Dogs will solicit play from crows with the canine play bow.

The mutual ability of corvids and canids to understand each other's intentions may be ancient. Wolf expert David Mech reported on play between ravens and wolves in Minnesota. Ravens routinely pinch wolfs' tails and are often seen taunting wolves into a game of tag. Flying low over a traveling wolf pack, ravens incite them to leap or follow, and when provoking a resting wolf on the ground, ravens often stay just out of reach, stimulating a lunge or pounce. Rarely do these risky games end in a raven death.

Tony's raven, Macaw, would often treat the family dog as a wild bird treats a wolf. Macaw occasionally joined their young family and husky dog, Quinn, as they lounged about the yard in the summer sun. The raven was indifferent to their infant twin daughters on the blanket but routinely yanked on the sleeping dog's tail. Jolted from his slumber, Quinn would leap up to chase the fleeing bird up and down the creek, their croaks and yelps audible. Five minutes later the two would return to the yard exhausted, with the dog returning to its nap and the raven settling in on our porch. Their "game" continued through the summer months.

Macaw and Quinn enjoyed playing together.

Each case of a crow or raven bonding with a dog or cat is unique, but there are common themes. First, the corvids were loners. For unknown reasons, their typical social partners, siblings and parents, were not around. Without the typical social bonds that they would have formed among the nuclear family, they may have been highly motivated to find a social partner. Second, the corvids, and in some cases the domestic pets, were young. Play is a strong motivator for a young animal, be it a kitten, pup, or fledgling crow. A young bird is also inexperienced and apparently trusting. Young corvids instinctively freeze or seek cover when they spot a hawk or owl, but this reflexive response to aerial predators does not extend to domestic cats and dogs. This may be a relic of the natural dependency that corvid scavengers have on mammalian predators or a reflection of the ability of birds to more easily escape terrestrial versus aerial predators. In either case, it seems that young corvids engage domestic pets, and possibly wolves and other natural associates, with a tit-for-tat strategy; if the predator offers play and companionship, then the crow stays; if aggression, it flees.

The third and perhaps most important feature shared by each interspecies association is that the corvid (and probably the other species) gains a reward for its efforts. In this way, the tendency of young corvids to trust potentially dangerous animals is consistent with their interactions with people. Just as Al, the young, lone raven adopted us in the wilderness of British Columbia, a young raven adopted Mary and Jim Ronback in Tsawwassen, British Columbia. Ravens are not unique in this respect; young lone crows have adopted people too. In all cases, including our own, the young, curious birds were fearless. They landed on us, walked and perched calmly around us, and ate what we offered. Each was consistently rewarded with food.

Like food, companionship and play are rewarding to a social animal. So, while rewards may be paid out in substantially different currencies—energy, a social bond, experience, or knowledge—they reinforce behavior through similar action in the brain. There, dopamine, hormones, and opioids motivate and reward the involved animals' behaviors. The trust afforded strangers by ravens and the delicate play between a cat and a crow are testament to the influence wielded by the brain's chemicals.

Social isolation is a powerful motivator. Alone, a young animal gives distress calls and seeks companionship, even if it is risky. When a social partner, typically a parent or sibling, is found, the cries dissipate and calm returns. Acquiring a social partner after isolation causes endorphins to be released that then bind to neurons of the septum, striatum, preoptic area, thalamus, amygdala, and hypothalamus. This opioid reward replaces emotional distress with comfort and pleasure. Endorphins are important to the organization of social behaviors like affiliation just as they are to other reward-seeking behaviors like foraging, sex, and play. A young lone crow or raven would be highly motivated to alleviate its stress by finding a buddy.

There are many reasons why a young bird might seek companionship from another species. A lonely bird might be evicted from its native ground by defensive parents or orphaned even. Perhaps all of the members of its species have been too aggressive or the bird's subservient social status made it difficult to join a flock. It is even possible that the young crows we learned about had abnormally low levels of vasotocin or mesotocin in their brains, which would interfere with their usually accurate species-recognition ability. But we think it is likely that the young birds that bond to other species are normal explorers, free for the first time from the familiar and comforting natal territory. Their natural curiosity guides them. As they encounter their world, they test it and, as social animals, seek companionship.

Testing the reactions of other species may be quite normal for a young crow. Most such tests would teach the bird what to avoid, but this is not absolute. Some cats, dogs, and people may reciprocate a corvid's curious approach and welcome the company. A social bond is a two-way adventure, after all—both partners need to accept and foster the relationship if it is to endure. A crow that behaves in a trusting or curious fashion might encounter restrained or playful behavior from a dog, cat, or person. If cat and crow both, for example, are motivated to seek companionship, then both will be rewarded by it and behave in an appropriate way. When the search for a companion is reciprocated, the young crow (and cat, dog, or person) is hooked. The reward circuit of their brains provides a powerful opioid fix that keeps them coming back for more. They are interspecific social junkies.

Once a young crow is attracted to another species, the brain regions that convey social meaning and motivation sculpt a partnership. Social meaning is encoded in the linkages among neurons in the amygdala, forebrain, and hippocampus that are sensitive to experience and ongoing rewards. The meaning of a partnership affects and is affected by the motivation and reward circuits between the nucleus accumbens and other areas of the midbrain and forebrain. Dopamine released in response to another species' offer of food, play, or companionship binds to the receptors in the nucleus accumbens motivating the formation of a partnership. Receptivity to another's advances would also be increased by dopamine. As two individuals interact over time, more receptors bind the dopamine, which finalizes a partnership between species just as the binding of dopamine in a prairie vole's brain seals a monogamous pair bond. The hormones mesotocin and vasotocin appear to motivate animals to seek social rewards, rather than, say, sweet foods, by linking the pleasure of companionship—nuzzling, play, and safety—to the motivation of dopamine and the reward of endorphins.

The mechanisms that motivate learning about social partners are built on well-understood reward-based motivation circuits in the brain, but at least one alternative hypothesis that involves a different form of learning may be at work. Young animals rapidly form lifelong memories that guide their immediate recognition of family members, the recognition of their species, and the future selection of a mate. This specialized learning or imprinting in birds involves synapses formed in the central part of the forebrain, just below the hippocampus. Filial imprinting—learning to recognize parents and one's own species—happens in most birds within a few hours to days after fledging from the nest. Could crows befriending other species have imprinted incorrectly? It is possible, but the social partnerships we investigated formed well after the age at which imprinted memories should have been formed. We think that the crows and ravens who bond with other species still know who they are. In fact, those that originate from the wild eventually leave their unusual friends and reintegrate with flocks of their own species. Imprinted identity may have more to do with ending, rather than beginning, the bonds formed between species. In autumn when wild flocks of crows and ravens gather, or in the following spring when hormones motivate breeding behavior, imprinted memories may call a growing crow back to its own society.

All animals take risks, albeit calculated ones that are evaluated by natural selection. The differences in the degree to which they risk death or injury may depend on the species, the individual, its sex, or its age. Corvids, it seems, can afford to be gamblers. Taking a risk may place a young bird in a beneficial relationship with other species, and it often pushes a seasoned bird to confront a potential predator. Throughout the animal kingdom, confrontation is typical of the ways in which species interact. Members of different species naturally eat one another. Others directly or indirectly compete. Mutually beneficial symbioses are mostly rooted in practicality, rather than pure enjoyment. Playful friendships between species are less common, but corvid play can create a condition that is nonthreatening and allows the birds to explore other potential benefits from associating with another species.

Considering the curious nature of a crow and the tolerance of our domesticated associates, it is perhaps not at all surprising to see a kitten and crow or dog and raven share good times. Even the approach of a wild crow or raven to a tolerant human is believable. We would find it surprising not to discover occasional young animals from different species enjoying one another's company. The chemistry of their brains, especially the pleasurable sensations of opioids that reward play and companionship, make the joy of such behavior addictive to social animals. Most cravings are satisfied within a species, but not always. Befriending a member of another species can be mutually rewarding, and for this reason it is a risk worth taking. To crows it may be simply irresistible.

9

Awareness

Hungry magpies learn to ring doorbells to solicit food.

In Sweden, Inda Drakborg heard the doorbell ring, which startled her from her morning routine. She hadn't noticed anyone in the driveway. The doorbell was located at the formal front door and was used only a few times each year; friends and deliveries came to the side door. As she opened the dark wooden portal, she was met by only the blinding morning sun. A bit puzzled, Inda went back to her work, only to be disturbed again by the bell. Same story—no one outside. On the third ring, Inda saw a familiar shadow fly from the door. Across the road, convened atop the neighbor's roof, was a small band of watchful magpies. So she offered them a scrap of chicken, something she had done infrequently over the last few months. Having felt sorry for the birds during the harsh winter, Mrs. Drakborg had been feeding them kitchen scraps in her back garden. The birds livened the snowscape with their flashy plumage and pleasant chatter, occasionally approaching the windows.

Yet that day, one of the birds had been especially innovative, ringing the doorbell as a child might do in search of a treat on Halloween. Her curiosity piqued, Inda watched the birds carefully throughout the long Scandinavian winter. When the bell rang, she fed the birds. Finally, she was able to see a bold black-and-white magpie cling to the brass lion's-head doorbell and push hard enough on the gilded beast's tongue to set off the bell.

Unknowingly, Mrs. Drakborg was shaping her magpies' behavior, rewarding their begging actions with a bit of bread or meat. The more she rewarded them, the more elaborate their begging. The birds quickly associated food with Inda, and Inda with the bell. And they generalized the function of the doorbell. They had little or no opportunity to see the front doorbell in action, but they could daily observe visitors ring the similar lion's-head device at the side door. Perhaps the magpies used the front door to avoid a surprising and possibly dangerous encounter with someone other than Inda.

These magpies, close relatives of brainy crows and ravens, were smart, but not blindly obedient. They were precise in their discrimination, adjusting their behavior according to whom they encountered on the other side of the door. Although Inda fed and encouraged the birds, her husband, Abbe, did just the opposite. He complained loudly about their presence; he enjoyed feeding sparrows, but his culture had taught him to discourage magpies. To him they were pests, and he shooed the birds at every chance, never fed them, and at one point pretended to hurl a rock at them. Almost immediately the birds took offense. They rang the bell only when Inda was home. Since Abbe was now seen as a source of wrath rather than sustenance, the birds took another course of action and whitewashed his Mercedes at every chance. The feathered bombardiers were persistent and precise, hitting only the driver's-side windshield with runny splashes of excrement. Before he could drive to work every day, Abbe had to wash the rank magpie crap from his car. Incensed, he parked his car in his wife's usual spot, but her Volvo remained untouched. Inda found this all hilarious, which might explain why she started to feed the gulls who now frequent their new home in Malmö.

These magpies recognized the Drakborgs and their autos and used this knowledge for their own benefit. Reports from around the world suggest that crows quickly recognize people who regularly feed them. In Seattle, we can spot people feeding crows from quite a distance—a black cloud swarms behind the feeder, gathering peanuts or pet chow, while the hungry fly ahead, perching in anticipation of the next dole. Larry Makinson from Bandon, Oregon, is so thrilled by his flock's response to peanuts that he refers to it as "taking the crows for a walk." In Minnesota, James Anderson regularly feeds almonds and peanuts to crows from his patio. One obsessed bird would pound on the glass door to demand his daily ration. Afraid for the bird's health, James stopped feeding, but it took nearly a year before the crow stopped banging. Shelly and Jim Leonard fed their neighborhood crows for eighteen months until a single bird began to wait for them on their porch. Maybe it was the special shrimp treat the Leonards offered, but on an otherwise typical September day, Jim entered his kitchen to find a crow inside the house working its way toward the dog's food. The bold bird quickly retreated out the back door.

Crows also quickly recognize cars. Phyllis Alverdes has been feeding crows for twelve years from her silver Toyota hatchback. As John watched her pull into a shopping center lot full of cars one morning, a swarm of crows descended to surround her. They routinely escort Phyllis as she drives several miles from her house to work, which caused a problem for Phyllis as a school-bus driver. Afraid that the birds would start terrorizing the bus and the kids, Phyllis's supervisor banned her from parking within a hundred yards of the bus lot.

In contrast, dangerous cars are remembered as threats. Ornithologist Carl Marti was observing a rural crow nest in Colorado. Periodically, he climbed the nest tree to check on the developing chicks. After his first visit to the nest, the adults recognized his car and started yelling as soon as they spotted it. John had a similar experience with ravens in Idaho. He drove a gray Toyota pickup to a nest every three days so that he could catalog the growing brood. After a single visit, the defensive adults would meet him nearly a mile from the nest and fly directly over his truck, scolding him. One day he drove an identical make of truck that was white, and the birds ignored him until he stopped at the nest.

Even dogs and cats gain a reputation with the local crows. Professor Prisca Cushman was walking to work in Minneapolis with her dog in 2006 when it suddenly discovered a young crow in a bush. Each day for the next week a mob of crows flocked around her dog whenever it ventured across campus. Prisca felt compelled to leave the dog home, after which *she* became the object of the crows' wrath. For weeks the birds would gather around and scold her, even gathering outside of her second-story office window to continue the vocal assault. In Virginia, Jingle Ruppert tested the ability of crows to distinguish between her two border collies. Her black-and-white collie ignored crows, but her tricolored collie always barked and chased crows from the yard. When Jingle walks her dogs around the neighborhood, the crows scold only the tricolored dog. When she walks only the black-and-white collie, the crows are mum.

Gene Carter, a Seattle resident, learned firsthand how tenacious a crow's vengeance can be. Finding a pair of crows approaching a robin's nest in his backyard, Gene decided to intervene. He pulled his black parka up above his head, like an outspread, upended cape and set about jumping and flapping toward the crows. They began to mob him but soon headed back toward the robin nest. Gene heaved a broom and black plastic garbage bag at them, scattering the crows. Each time the predators headed for the robins, he javelined the bag and broom contraption toward them. The robins survived, but the next morning one of the crows spied Gene fixing coffee in the kitchen. It began scolding. It followed Gene to the bus stop, still scolding. Every day for the next year Gene was escorted to and from the bus stop by an irate, vocal crow. In openings between trees, the crow would dive toward Gene, occasionally rapping him on the head, always amusing his wife and fellow commuters. If Gene stayed at home, the crow would maneuver around the house until it could see him through a window and scold. After a year Gene moved to a new house, twenty blocks away. He was so paranoid that the crow would follow him that he drove his belongings via a circuitous route to the new abode. Any sighting of a crow from the car would cause him to abort the trip. When the last load was secreted to the new house, Gene told his wife he was going back one more time to clean up a bit for the new tenants. He stayed in the empty house watching a portable television until three in the morning, so he was sure the crow was asleep before he made his final move.

We know how Gene felt. As we study corvids, especially when we need to catch them, we often feel scrutinized by them. We have never yet been shat upon, but others have. Lawrence Kilham was in Iceland trying to shoot ravens for a study when he fired a gun at a bird and missed. The bird was more accurate however when it released its own body-heated missile and scored a direct hit on Kilham's field garb.

One cool, early spring day, trapping crows in 2002 just outside of Seattle, Marco Restani and John hid their net gun in a small clump of ferns and grass, just as we did to catch ravens in the North Cascades. We baited the roadside with fresh Cheetos. Just after dawn, a murder of the wary birds hurried to our bait like a bus full of tourists to a five-star buffet. They pinched together, shoulders touching, beaks jabbing, squawking a bit, but mostly quiet with heads down and beaks dusted orange from the cheesy powder coating their food. The hungry birds seemed oblivious to anything but the fat-laden chow. We fired the gun, sending four heavy metal cylinders over and past them, so the trailing net ensnared the Cheetos and Cheetos eaters, then we raced to the net to secure the birds.

The first crow John held, measured, and surveyed still smelled of fake cheese as John slipped four rings around its polished, scaled legs. Ankle bracelets, three of plastic, one of silver—light blue over dark blue on the left; powder blue over a metal federal ID band on the right. To us this crow now became an individual that we could recognize, follow, and learn from. Over the coming years John and his students would look for his nests, handle his mate and kids, and learn his territory. Because John could now identify him, he paid special attention to him, directing scopes, binoculars, antennas, and microphones at him, stopping and staring whenever he saw the bird, and searching him out. Through this scrutiny John had an unprecedented glimpse into an individual crow's world. For seven years, it was a rare day that Light Blue, Dark Blue and John did not exchange glances. John looked at him with wonder, but the crow never forgave or forgot John. As soon as the bird saw John, or his truck, regardless of what he wore or with whom he traveled, the crow flew to the safety of a tall fir tree and cawed harshly. He followed as John walked, scolding from above, alerting other crows that danger was afoot. To him, John was no different from other local predators: coyotes, bobcats, raccoons, hawk, owls, and eagles.

Light Blue, Dark Blue was a neighbor, owning territory a few blocks from John's home, a neighborhood that became gentrified during the crow's life. The rapid subdivision of his once-forested land brought him new foods; lawns rich in worms and crane flies, roadkilled squirrels and opossums, and well-stocked bird feeders. But it also brought in a diversity of people. Some, like John, seemed always to be after him. Others, like Bill down the street and Carolyn around the corner, arranged their days so they could feed him rich mixes of peanuts, cat food, seeds, and suet. To survive in our world, this crow needed to learn our intentions and habits. He learned to exploit many and avoid some. He and his mate raised two or three fledglings every year, until the spring of 2009. To the northeast, a few strokes of a wing away, a new resident lured him to his yard and shot him. Curiosity and trust cost Light Blue, Dark Blue his life. This unpredictable threat that people pose favors discerning corvids over those that cannot tell us apart or remember our past actions.

It seems natural for wild animals to recognize food and predators, but how do they recognize a person? Might they just know cues to our identity—our habits, clothes, walking styles—or do they really know us as individuals? To test the recognition abilities of crows,

John hit upon the idea of changing his identity and seeing whether the crows could learn that his new identity was dangerous. Surfing the internet looking for disguises, he found Dick Cheney's face could be bought for $29.99. John ordered one mask, and a second similarly bald, but distinct face—a caveman mask. One would be used for trapping to test the crows' abilities to remember a dangerous face, while the other would remain neutral, never used for trapping, but rather to control for any potential issues crows might have with masks. John's students John Withey and Jeff Walls were willing to join the fun.

We used extraordinary masks of cavemen as we captured crows in Seattle as a means of testing their abilities to recognize human faces.

A couple of days before Valentine's Day 2006, students and professor donned grotesque masks—bold, heavily browed, reddish-orange cavemen—and captured seven crows on the University of Washington's campus. They tagged the ensnared crows with standard plastic and metal bracelets like those we had fit onto Light Blue, Dark Blue's legs and released them after only a few minutes. On Valentine's Day John slipped into his Dick Cheney face and strolled across campus looking for crows to record their reactions. He found nine birds, and while one seemed a bit anxious and flew off calling, the others basically ignored him. The students were more reactive, as being Dick Cheney on a liberal college campus wasn't easy, but from the crows' perspectives Dick was just an average Joe.

Two days later, John left the Cheney mask in the lab and morphed once again into the caveman. He stepped outside his office building at 11:07, eager to learn whether the crows would remember the face of the man who had captured them earlier in the week. At 11:15, he found a crow near the student union building and began to approach. Immediately the bird flew into a tree and gave a series of harsh calls, flicked its tail, and stared directly down at him. This scolding behavior, identical to how these rowdy birds typically address their natural predators, quickly attracted a second bird. The pair now cautiously eyed John and issued a real tongue lashing. The first scolding bird was unbanded—John had never even handled this aggressive beast. But the second bird wore bands, signaling that it had personally met the caveman a few days earlier. This bird had good reason to scold—the caveman was a proven threat. But the first bird could have known only secondhand about the dangerous caveman. Perhaps she had seen us catch and band her colleague. John continued his walk and in total encountered thirty-one crows, three of whom scolded him.

The first run of the experiment was a success. At least three birds recognized and harassed the dangerous caveman. In contrast, none responded to the caveman prior to trapping, and none responded to the "control" face of Dick Cheney, who had never directly participated in trapping. We repeated these initial tests with similar results over the next year. We even recruited other students to run the tests for us. We wanted to make sure it wasn't just our imagination or perhaps the way we approached the crows that made them scold the caveman and ignore the vice president. We set the students loose on campus with masks and notebooks. Their results confirmed ours in every aspect: the crows scolded the caveman, not Cheney; many of the scolding birds were unbanded; and it was the face that triggered the ire of the crows.

We have continued and expanded our initial investigations. In addition to the caveman on campus, we have now confirmed other crows' abilities to discriminate dangerous from neutral faces in four new settings. And we have done so using masks molded from our friends' faces—ordinary men and women faces much less distinct than the caveman's. In downtown Seattle for example, our friend Scott's face was used during trapping. As with our campus experiment, the local crows screamed, dove, and followed anyone wearing his mask while ignoring those wearing any of the other five masks. In rural Maltby, Vivian was the trapper. There she was scolded while Scott and the others were more or less ignored.

When encountering a single face, crows do occasionally scold a person who is not dangerous. It seems safer to cry wolf occasionally rather than ignore a real threat. But when we presented the crows with a choice, their ability to distinguish among people was uncanny. In this experiment two of us approached a crow while we each wore a different mask, one dangerous and the other neutral. As we neared the crow, we diverged in opposite directions for a while, then reconvened, and diverged again. As we paraded back and forth, invariably the crow lit out after the dangerous person, following him and letting the other masked, but harmless, person strut unscathed.

Crows may remember our facial features or perhaps have a simple signal—maybe a special call—for dangerous people in general, though the latter seems not to be the case. We recorded the voices of crows as they screamed at us and at hawks and raccoons and found no obvious differences in the calls to people generally or to dangerous people specifically. We know from other studies that corvid alarm calls indicate the caller's identity and often the degree of threat posed but not the specific identity of a predator. Siberian jays, for instance, adjust their alarm calls to encode the hunting behavior of hawks—moving versus perching versus attacking. Crows may do something similar; the intensity, duration, and pace of scolding indi-

After we trapped crows while we wore a particular mask, the birds quickly recognized us and scolded and mobbed us whenever they encountered us.

cate the degree of danger a predator poses. But this adjustment of scolding occurs whether in response to a hawk, coon, or caveman.

The lack of particular alarm calls for specific predators may have more to do with the ancient and innate nature of these vocalizations than with the inability to distinguish between predators. All vertebrates have vocal centers in their midbrains that, when stimulated, elicit innate calls—a human scream, a rooster's crow, or a corvid's alarm call. In birds this is the nucleus intercollicularis, which receives input from the forebrain song-learning circuits, the hypothalamus, and the nearby optic tectum (these regions of the brain are illustrated in the Appendix, pages 226–227).

The sight of a predator is projected from the eye onto the midbrain's optic tectum, like a movie on a theater screen. The neurons in the optic tectum are especially sensitive to the movement of objects; their activity rate parallels a predator's motion. The inner neurons of a bird's tectum respond not only to sight but also to sound and touch, so that they can integrate all of their senses to detect and track a dangerous object. Pigeons, for example, use sight and sound to recognize individual people, and this information is either integrated in the optic tectum or in the forebrain. When a crow sees a predator, a similar degree of coordination in the midbrain likely connects aggressive posturing within a mob—crows that are flicking wings and tails—to the alarm calls they are literally pumping forth. Stimulation of a crow's tectum affects the midbrain vocal center, which it surrounds, and then the muscles controlling the syrinx, breathing, and movement to produce a stereotyped posture and cry or scream.

In some songbirds, alarm calls may be learned, and in this way the calls come to label specific predators or to be incorporated as elements into complex songs. But this does not seem to be the case for crows. The crow's forebrain may be needed to identify an object as a predator and associate it with the production of innate, harsh cawing, but the brain of a crow is not used to devise unique calls for specific predators.

We suspect that when a crow sees a predator, its adrenal glands release corticosterone that binds to receptors on neurons in the brain stem and causes the rapid release of norepinephrine into the social brain network, particularly the amygdala. Especially, but not only, during the breeding season its gonads would also release androgen hormones—in males, testosterone. This hormone may influence less immediate mobbing reactions than do the rapidly acting chemicals like epinephrine. Focused by stress hormones, and aggressive on testosterone, the birds confront predators. Androgens, corticosterone, and epinephrine also bind to receptors in the midbrain vocal center, which primes these neurons to fire easily when stimulated by the sensory-informed, surrounding regions of the midbrain or forebrain. This may explain why mobbing intensity, even in many corvids, increases during the breeding season. Their aggression is fueled by hormones and finely targeted by experience and memory.

While the noise of a boisterous mob surrounding and diving at a malevolent person may involve reflexive, nonthinking activities commanded by a crow's midbrain, the decision to join a mob, the acquisition of knowledge concerning danger, and the ability to use this knowledge to recognize a person requires thought. The striatum, including the nucleus accumbens, of a crow is important in this respect, as it is in associative, reward-based learning in general. Here, many linkages among parts of the social brain network are forged under the influence of dopamine, opioids, and stress hormones. The scene perceived by a crow is assessed by its forebrain: the entopallium informs a bird about what it sees, the Wulst and hippocampus work together to determine where it is relative to known landmarks and other such aids, and the amygdala adds emotional context. Critical connections among neurons from all these places may occur in the nucleus accumbens or in the amygdala itself. Neurons in the striatum are informed by birds' senses and experiences and are shaped by emotion to produce reasoned actions like mobbing. What is additionally complex about associating a particular person with a reward, either positive or negative, is that the bird must first recognize the person.

Dogs, monkeys, sheep, honeybees, octopuses, pigeons, and many other animals recognize individual people. A face, like any other reliable cue, could trigger the release of dopamine and thereby motivate behavior. In humans, facial recognition, while involving many brain regions, is well understood. A core recognition system is dispersed among at least three sensory regions in our forebrains and networked with two extended systems that convey the history and emotional significance of the person. This network permits us to quickly decipher a person's traits, attitudes, emotions, and mental states and recall our past interactions with them. In total, ten or more distinct regions of the human brain act as a complex, interconnected network to analyze the identity and importance of each face we encounter. Initially and subconsciously, we assess the emotion, then the identity, of a face. Secondarily, we consciously consider the face. This integrated assessment of each face we meet allows us to act strategically in social situations and plan our actions to attain future goals. With a glimpse of a face, we understand something about the other's emotional state, intention, and the significance of the person to us. Could crows possibly know as much when they look at us?

We are far from having a complete answer to this question. Our work with masks proves that crows are aware of our prior behavior and reluctant to accept the possibility that we might change. Ravens' attention to the direction of our gaze proves that these animals also read our faces for subtler clues to our intentions. We have recently confirmed this in crows as well; looking directly at a bird causes it to flee sooner than does looking away from it. Whether birds understand the significance of our more subtle facial expressions remains unstudied. The birds we observed did not seem to distinguish between a smiling and a scowling face. The tendency to smile or frown may not reliably cue a crow, but we suspect the crows do notice the differences. This is in part because of the distinct way in which birds perceive complex patterns. Primates, including humans, are especially attentive to the overall configuration of the face, not its component parts. Birds appear most attentive to the parts and are therefore better able to recognize a face even if its parts are in the wrong places or in changed posture. Our crows are like pigeons in having little difficulty spotting a known face even when it is inverted. We created an upside down caveman mask as part of our experiments. While annoying to wear, the inverted face was consistently and strongly mobbed by the campus crows. They recognized the parts as being the caveman but also understood the distortion; often when seeing one of us in this mask, an alert crow would turn its head upside down and eye us as if to say: *"You're not fooling me!"* Then it would right itself and scold us for all it was worth. This ability to see the key features of a face despite their orientation and arrangement may help crows pick out subtle but important differences in our expressions. Looks like we might need some new masks.

While observing behavior tells us something about how crows view us, it tells us very little about how their brains reacted when they saw us. That all changed when John had the good fortune of meeting a few new colleagues at the University of Washington Medical Center. Dr. Donna Cross, Dr. Robert Miyaoka, Barbara Lewellen, and Greg Garwin specialize in the imaging of animal subjects. They routinely scan the bodies and brains of monkeys, rats, and mice to understand brain function, cancer, posttraumatic stress, autism, and other human afflictions. When Heather Cornell, a student who conducted our studies of human recognition, and John met Donna,

Birds seem to recognize the individual parts of a face. Because of this, the face of an enemy does not fool them, even if it is presented inverted. Sometimes the curious birds even rotated their own heads to confirm the dangerous identity.

Robert, Barbara, and Greg, they had never peeked inside the brain of a bird. But they were game to try.

We decided to make the first attempt to measure how a crow's brain reacted to the sight of a dangerous person. Heather and John caught thirteen crows wearing one of our custom-made, realistic masks. As our birds acclimated to their individual, outdoor flight cages, we fed and tended them while wearing a different mask. By keeping their contact with other people to a minimum over the next few weeks, we would be able to compare their reactions when they glimpsed a familiar-but-dangerous versus a familiar-but-caring face. We were ready to look at their brains in action. The procedure Donna and her team proposed is called PET (positron emission tomography) imaging, widely used in people and animals and in order to visualize where the brain was active in the half hour or so before the scan is done. This ability to look back at a crow's mental activity is important because during the scan the bird must hold still, which requires anesthesia. We also used the more familiar scanning technique, MRI, but because this technique scans what is currently happening, not what has happened, we used it only to get a detailed look at the structure, not function, of the crow's brain.

Here is how we scanned the crows' brains: We brought a test crow up to Robert's lab and allowed it to get used to a new, smaller cage for an evening. The next morning John reached through a curtain, grabbed the crow, blindfolded it, and gave it a small injection of radioactively labeled glucose. John then put the bird back, playing soothing crow calls for a few minutes, and then Heather and he removed the curtain to allow the bird to see their faces. Some of the birds saw them wearing the mask worn when they were captured, some saw the mask worn when they fed them and cleaned their outdoor cages, and a few saw no person, only the room. After fifteen minutes of on-and-off viewing, Greg and Barbara anesthetized the bird, wrapped it in a warm blanket, monitored its vitals, and slid it into the PET scanner. As the crow slept, the scanner mapped the occurrence of the radioactive tracer within its brain. As the bird was viewing Heather and John, its whirring brain was demanding energy that was supplied in part by the labeled glucose they had slipped it. The most active parts of the brain would require the most glucose and therefore receive the greatest label; less would accumulate where the brain was less active. The scanner, like a 3-D X-ray machine, would paint us a picture of the tracer's deposition. In an hour we would know how the crow was thinking. In a day, after clearing its body of the tracer, we could free the crow and let it resume its life in the wild.

As the first crow awakened from his isoflurane-induced sleep,

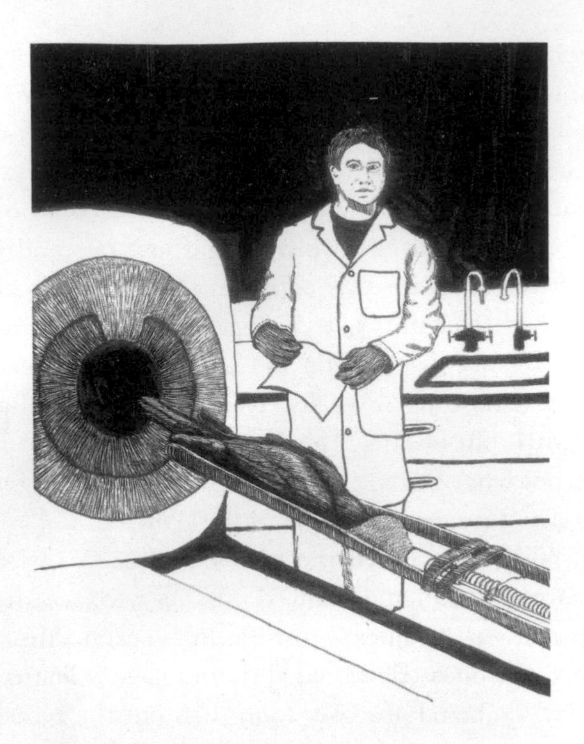

We used PET imaging to scan the crow brain.

Robert was already busy reconstructing an image of the bird's brain. Amazingly, we had a computer rendition of the crow's brain that we could explore bit by bit. Right away we saw the eye's retinas, the thalamus, and the entopallium blazing with activity—this was the trace of visual information streaming into the crow. There was also activity scattered throughout the forebrain. Our crow had taken in the scene. But would the brain reflect anything about whom the crow saw, or would it just reflect fear of a strange place, a sudden grab, and a shot in the belly? Donna next worked on the images to compare, point by point, the brains of crows that had viewed dangerous, caring, and no faces.

The results were striking. As in human images, we saw a complex network of brain regions respond to our presence. Sensory areas in visual pathways translated sight into neural activity. The integrative nidopallium and mesopallium, and the associative striatum were

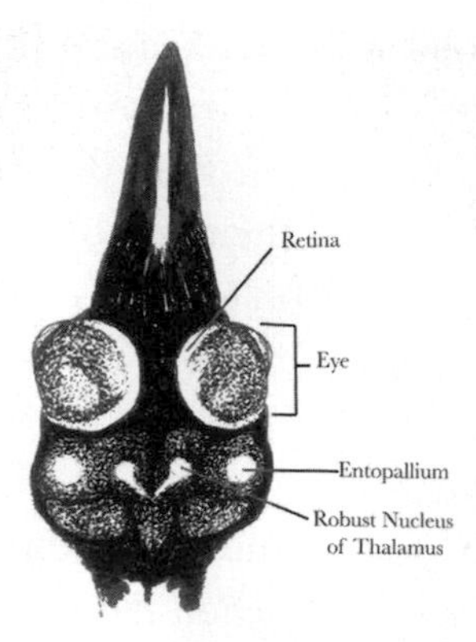

Horizontal
View

A raw image from PET scanning a crow brain shows the energy used by the eye (retina and choroid), the thalamus, and the forebrain (entopallium). Much of the rest of the brain, including emotional and memory centers, is also active.

active, as expected if our crows were evaluating their visual experience in the context of memory. When looking at a person, crows used one side of their brains more than the other; their right forebrains were especially active. And some areas appeared especially tuned to the dangerous face. When viewing a dangerous face, our crows used their nidopallium, arcopallium, amygdala, and areas in their thalamus and brainstem known to be important to fear responses. This reaction was remarkably similar to that of a person who views a dangerous situation. Our crows even relied mostly on the right hemisphere of their brains, just like people do in fearful settings.

The activity in the brain of a crow who looked upon a caring person was quite different from that of a crow who saw a dangerous person. Upon seeing a caring face, the preoptic area and striatum of the brain were most active. These regions are known to be part of the social brain network stimulated during social interactions, where their activity indicates a bird's hunger and its attention to learned associations. This suggests that crows perceived the association they learned between food and their human caretakers. Again, our crows even varied the use of their two brain hemispheres, exactly as do humans. Instead of using their right brain, as was the case when seeing danger, now they used their left brain. Clearly, as with humans, crows pay attention to peoples' faces and integrate what they see with what they remember and feel, using a complex neural circuit to evaluate each of us.

As Heather and John watched each bird watch their masked faces, they noticed that some froze and fixed them in their gaze, while others moved their heads from side to side and frequently blinked their conspicuous, white eyelids. To us the birds who froze at our sight looked scared just as they did when we first captured them, while those regularly moving and blinking seemed more relaxed. Mostly, but not exclusively, the scared birds were looking at the dangerous mask, and relaxed birds were eyeing the familiar, neutral mask. The activity of the brain region including the amygdala of each bird supported our intuition. The number of times a bird blinked per minute was nearly perfectly associated with the relative activity of its amygdala. Just as the amygdala of a quail governs over its response to freeze or flush at the sight of danger, so, too, does the crow's amygdala. The less a crow blinked—a sign of fear—the more active was its amygdala. The strong connections among the motivational and associative centers of the brain, including the preoptic area and striatum, and the amygdala, which is part of the emotional brain, sometimes called the limbic system, enables crows to use memory and emotion to guide their actions. Our crows' emotions could be detected in their eyes and within their brains.

Convinced that crows know the face of danger and not just the strangeness of the caveman or a mask, we continued to dress up and walk around campus. On August 31, 2009, three and a half years after trapping, John was wearing the caveman mask, doing his rounds. Of thirty-eight crows he encountered, two-thirds scolded him. He was hounded the entire time he wore the mask; even birds of other species—jays, nuthatches, and chickadees—scolded him. A few months later, on the occasion of the four-year anniversary of trapping, the response was a bit more muted, as he encountered only sixteen birds, seven of which scolded. Only one small mob of three birds paid him any significant attention. Is the caveman finally being forgiven or just forgotten?

On February 15, 2011, five years since the caveman did his dirty deed, and the first time John had worn the caveman mask on campus in one year, he found thirty-seven crows. Three out of every four he encountered, twenty-eight in total, scolded him. Within a few minutes of starting the test, John was surrounded by an aggressive mob of twelve crows, none of which appeared to have been banded previously. The word was spreading like a bad joke; everyone knew the punch line. And yet John had done nothing to the campus crows for years. Wow, they really hold a grudge.

Not only was the crows' hatred of the caveman persistent, it was getting worse with time. In the five years since we trapped on campus, the number of birds scolding the caveman on a typical walk has increased threefold. And the vast majority of those who berate the Neanderthal were never even touched by him. By looking closely at those mobbing the caveman, we have identified four ways that crows come to know his face: First, a few learn by direct experience; those we captured are nearly perfect in scolding the caveman, and only the caveman, in the future. Second, a few learn by observing us capture their flock mates; those that we missed with our nets circled our trapping site, scolding and watching, learning about the caveman just as we suspect other crows learn from the scene of a comrade's death. Third, fledgling crows learn from their parents. A few summers ago we closely watched twelve crow families learn about the caveman and other faces we used for trapping. When the parents scolded us, so, too, did the young fledglings. And, most important, when the parents were not around, those observant fledglings scolded the dangerous face without parental coaching or coaxing.

Finally, most crows learn about us from their peers. They watched and joined in with other knowledgeable birds who scolded us. Scolding is contagious, so when one bird scolds, all the others within earshot fly in to join the mob. Exposure to a mob doubled the chances that crows we had never captured, who were alone when we spotted them, would scold us. So crows learn about dangerous faces in many ways, some from individual experience, but most from tapping into public information. This latter mode—social learning—is unique among animals and cognitively challenging. Learning from observing others enables traditions—what we consider culture—to develop in animals like crows.

Some nerve cells may have evolved to support social learning. In a monkey, for example, a patch of neurons in the forebrain fires when the primate grasps an object and when it observes a person or another monkey grasp the object. An object alone, or an outstretched hand alone, is insufficient to excite the neurons. Neurons with both sensory (visual in this case) and muscular (motor) function are called mirror neurons. Their pattern of activity translates the information gained by watching another demonstrate into the motor commands needed to perform the modeled action. The direct neural translation from observation to action allows observers to understand a demonstration without complex thinking. They see it and do it, almost reflexively. We have mirror neurons in many parts of our brains that help us to understand the emotional expression of others as well as the goals and intentions behind the actions we observe. These neurons also help us to translate what we hear into what we say. Mirror neurons occur in many of the same areas in monkey and human brains that are important to face recognition. In songbirds, neurons with mirror properties have just been discovered in the HVC, an area that may simplify a young bird's task of learning his father's or another tutor's song.

Although mirror neurons in crows have not been investigated, we suspect corvids' abilities to imitate human voice are supported by mirror neurons in the HVC just as such neurons support more traditional song learning. We expect that mirror neurons occur elsewhere in the crow forebrain. Perhaps the regions that inform muscles also fire when magpies see humans ring doorbells, when New Caledonian crows watch tools being made and used, or when American crows watch their parents or peers scold a new threat. Copying complex play maneuvers would be greatly simplified if the neurons that encode action also responded to observation.

Our crows learn so efficiently by observation that we are convinced they will never forget the caveman's face. Every test we conduct only revives a fading memory and serves as a teaching moment for the ignorant. If crows don't forget the caveman's transgressions, will they forgive them? It's unlikely, since they don't forgive or forget even crimes less severe than our capturing and tagging them. Katrina Anderson found a dead crow in her Massachusetts neighborhood. While about a dozen crows scolded, she wrapped the deceased in newspaper and carried it a few houses away for burial. During the process Katrina was constantly observed by the mob of now-silent mourners. The crows associated Katrina with the death of a flock member, and the next morning, and every day for the next month, when she left her house, crows flew above her like a swirling black cloud and scolded her loudly. One bird seemed especially interested, calling to her in her yard and peeping at her through the windows. Randy Gerber had a similar experience in the spring of 1988 in Seattle. He picked up a young fledgling with the idea he would adopt it as a pet. The next day his wife persuaded him to release the bird, whereupon the local crows mobbed him vigorously. They continued to do so daily for the next three months, frequently diving toward his head.

Crows might take some active persuasion to forgive the caveman. Donna Barr rescued a grounded fledgling from a busy intersection, placing it in a high, safe tree crotch as the parents looked on. She was never mobbed. At a remote landfill in the Haida Gwaii archipelago off British Columbia, Bill Romaniuk was preparing to photograph eagles and ravens. As he arrived, the birds were visibly irate; a flock of ravens scolded and dived at him the instant he got out of his truck. He saw something bloody flopping on the ground—a raven had gotten trapped within a deer's rib cage that had been discarded by a hunter. Its wing was hopelessly tangled, so Bill opened his knife and started to cut away the bony prison bars. The hundred or more ravens that had scolded Bill were silent, perched on stumps, hummocks, and in the trees as he worked. Bill shook the bird from its skeletal cage and left. The attending birds did not mob him, and Bill was never mobbed by ravens again. As he worked the rest of the summer among ravens, he seemed to have a special status. While his colleagues in forest restoration often had their lunch pilfered by local ravens, Bill never lost a single chip, cookie, or sandwich. The entire forest crew often stashed their lunches side by side, yet Bill's was spared; even when left out in the open it was untouched. Bill was impressed, so when his son was born three years later, he named him Raven.

Ted Hayes and Nancy Kool also transitioned from enemies to friends of local crows. Each got on the bad side of different crows by passing too near a nest or fledgling. Nancy's dogs also seemed to threaten the crows. Tired of daily mobbing, Ted began to feed the crows peanuts. After about a week, the mobbing subsided. Nancy thwarted a sneak attack by a cat on a young crow and moved the baby to the safety of her garage roof. Nancy even gave it a can of cat food, which the parents parceled out among themselves and their fledgling. The crows stopped mobbing Nancy after this kind act, but they did toss the empty can of cat food toward her the next morning. It rolled off the roof and landed right at her feet.

That crows' memories fade slowly is consistent with what we know about how animals move past fearful experiences. Emotionally charged memories are rapidly acquired and longstanding. Fear tends to be especially persistent and is not forgotten, but it can be overcome by forming new memories. In laboratory settings, for example, rats quickly learn to fear a tone if it has reliably signaled an impending, mild shock. When the shock no longer accompanies the tone, the rats slowly stop fearing the once-terrifying sound. Neurons deep in the central part of their forebrains (the medial prefrontal cortex) are activated when the expected shock does not occur. This part of the cortex stimulates particular neurons in the rat amygdala known as intercalated cells, which then stifle further signaling of fear. A new memory has been formed—encoded as an activated neural circuit involving the forebrain and suppressive cells in the amygdala—that inhibits the fearful response previously associated with the tone. If we think of a train heading down a track, it is as if a new switch has rerouted the train in light of new information, and, as this new track is increasingly used, the old way is rendered obsolete.

Reduced stress and the resulting low levels of corticosterone may be important in overcoming fear. When little corticosterone circulates in the brain, this stress hormone preferentially binds to special receptors on neurons known as mineralocorticoid receptors (MR) and in so doing primes neurons for easy firing. Binding of corticosterone to these receptors in the medial prefrontal cortex, amygdala, and hippocampus would ready neurons to quickly learn that a particular cue no longer indicates danger. The effect of corticosterone on learning has been demonstrated: rats that receive low doses of corticosterone in their hippocampus more quickly learn to ignore formerly fearful tones.

So, feeding or demonstrating kindness toward a crow might really be the key to winning it over. Doing so reduces a bird's stress, which promotes the binding of low levels of corticosterone to MR receptors and readies neurons throughout the brain to quickly react to new experiences. New memories of helpful people are formed, and memories of once-feared people are suppressed.

The discriminating nature of wild crows forces us to consider our actions. Nature is watching. And some of her flock form lasting impressions that accurately reflect our best and worst behavior. Crows know us and are not afraid to voice their opinions. Knowing us, they also may know themselves as self-aware beings. They know who they are and how they fit into society. There is little evidence of self-awareness in animals other than apes. Not even chimpanzees, our closest relative, are known to be capable of it. The standard test for determining self-awareness in nonhuman animals is to present an animal with a mirror and study its response.

Invariably, when an animal first encounters its mirrored reflection, it responds as if a challenger, aid giver, or mate is present. When Al the raven joined us in British Columbia, we leaned a mirror against the picnic table and waited for his response. Al immediately became aggressive. He puffed out his throat plumes, raised his ear tufts, and while bowing, spread his wings and croaked hoarsely at the two-dimensional intruder. Of course the reflection returned every threat Al poured forth, so our young raven did not back down. He stepped from side to side and peaked behind the mirror for a

Al, the young raven, reacts to his reflection in the mirror.

better angle on his adversary. Convinced Al was not self-aware, we removed the mirror. We should have given him a second or third chance. Those animals that do recognize themselves do not do so immediately. Magpies, for example, are the only bird to have passed the mirror test. Three of five birds repeatedly tested eventually investigated a colored spot drawn on their feathers and visible only in the reflection. Upon seeing their surprisingly spotted throats—they had been marked while under the spell of anesthesia—they tried to scratch or rub the mark away. The performance of magpies equals that of the great apes. After that first sighting of himself, we bet Al would have recognized subsequent encounters with his reflection.

Corvids not only know us but may also know themselves. One day, perhaps, we will know how they perceive their lives.

10

Reconsidering the Crow

Crows form strong relationships with people. Here a pet crow takes his daily ride to school.

THE ACCOUNTS from our experiences, the careful experiments of many others, and the striking observations from citizen scientists have managed to paint a fantastic picture of the clever, opportunistic, social, and associative learner that is the crow. What at first glance is a common, ordinary animal we know now to be an extraordinary one. Much more than we realized is within the physical and mental grasp of birds in the family Corvidae. We are only just beginning to understand how the brain of a bird generally affects behavior, but we do know that corvids learn, remember, and think in much the same way as do other songbirds and mammals, including humans. Our neurons, spinal cords, hindbrains, and midbrains work to sense, transmit, and relay information and respond to our environments. Our forebrains are drastically distinct—assembled from reptilian bases but according to different blueprints—and built from some of the same embryonic tissues by the same chemical processes into functional, if not homologous, equivalents. The crow receives a vast array of sensory information that makes its way up to an extensive, yet modular, pallial forebrain where it is integrated, organized, stored, retrieved, and reconsidered. This information is employed both consciously and subconsciously to move muscles to the same end as does information in our own interconnected neocortex.

The impressive brains of corvids and humans enable long-living, social, adaptable animals to solve their problems in similar ways. We have sought to demonstrate that, in spite of our different cerebral structures, corvids nevertheless share many of the same attributes we claim for ourselves. Like the crow, we test our environments, observe the responses to ourselves and others, accumulate these experiences over our lifetimes, and use this knowledge for personal gain. Humans share knowledge in a variety of ways that a crow cannot—such as the words you are now reading—and we often act for communal or less direct gain than a wild animal has to in order to survive. But our basic approach to life—to investigate, try, observe, consider, remember, revise—is not all that different from what we believe is happening within the brain of a crow.

Crows clearly are exceptional at tuning in to associations between their environment and the dangers and prizes embedded within. They learn quickly from the errors and rewards of their trials. The associations between the features of their environment and the cost or benefit of their actions are codified in synapses among neurons from distinct brain regions. As with us, these neural associations are refined by a diversity of neurotransmitters, fostered by hormones, motivated by dopamine, rewarded by opioids, and tempered by context and emotion. Much of what crows and people do, they do because of these ancient chemicals that influence their and our central nervous systems. Our shared heritage as vertebrates is strong. We are both successful in part because we exploit, experiment, invent, innovate, learn, remember, and share.

Birds and mammals set off on parallel evolutionary adventures that have been aided by our distinct yet complex brains. Complex brains are not needed for success; insects, bacteria, and plants are more numerous and diverse than birds and mammals. But because of our mental complexity, we have traveled a different route. And not all birds and mammals have traveled the same route. Some, typically those that have specialized on a few resources, have remained relatively small brained. Others, like humans and many corvids, exploit a great diversity of resources (for example, the foods we eat). It is this adaptable, or "generalist," lifestyle that is often practiced by animals with large brains.

The cognitive power of a large brain, especially one possessing an expanded forebrain, supports the innovations that are necessary for a general lifestyle. Among birds, with the exception of a few of the large parrots, corvids have the largest relative brain size. Crows and ravens in particular belong to the most innovative genus of all birds—they employ the greatest variety of maneuvers to obtain food, and they have the strongest propensity to use tools and to do so in different ways than any other group of birds. The New Caledonian crow in particular is a poster child of crow mentality; of all birds, it has the largest brain relative to its body size and is unparalleled in its ability to manufacture and use tools. Perhaps we will learn that we share substantial characteristics with fish, salamanders, or lizards, but for now it seems that the greatly expanded forebrains of birds and mammals rule the mental day.

Large brains, while costly to fuel, benefit those they endow. Keeping track of interactions with other individuals may be a key advantage of maintaining a large cerebrum. Sociality increases food-finding ability, predator detection, and communal activities such as mobbing dangers, learning from others, and subdividing labor. Without the ability to recognize and remember the biographic histories of colleagues, be they feathered or furred, these benefits are reduced. The ability to live in social groups and learn from one another almost certainly increases the survivorship of corvids relative to other birds. It is not unusual for individual crows, jays, ravens, magpies, and other corvids to live for decades in the wild. In fact, corvids are among the most long-lived of all birds. In part, this longevity derives from their ability to figure out how to survive periods of climatic stress that affect food supplies and their capacity to reproduce. The innovative and social nature of corvids will also likely serve them well in the future. Animals that innovate and learn about inventions and discoveries by observing others have higher rates of morphological evolution. As species go, corvids will continue to evolve—mentally and physically—and amaze and challenge us well into the future.

The large brain of birds and mammals allows its owner to bounce ideas back and forth among its cerebral regions. Thoughts from the forebrain not only flex muscles, but they can also be compared to newly acquired sensations and be revisited and revised while asleep. This breakthrough allows birds and mammals to reconsider and edit their experiences and nuance their actions. This mental ability to know what one did, and then adjust and embellish it with what one later realizes to be useful, seems to us to be the root of complex cognitive activity. Such cognition depends on an ability to hold some experiences in memory and use them to adjust future action. We certainly saw our raven friend Hitchcock modify his view of windshield wipers in light of his experience and the new information it provided.

Having a sense of the past and the future, rather than only the present, allows birds and mammals to plan, change, and weigh consequences, and to refrain from unproductive actions. This is what neuroscientists consider the "executive function" of the brain, and while it remains one of the great mysteries of neuroscience to understand how it works, we can see birds and mammals using executive action to organize their behaviors. No single part of the brain is the sole executor. The prefrontal cortex of mammals and the back portions of the nidopallium of birds (NCL or CDL) play a role, but many regions surely interact to sequence and coordinate the translation of sensation into action.

Picturing the present in one's mind is a basic level of consciousness that most researchers ascribe to all vertebrates, even some invertebrates. Making executive decisions, being self-aware, understanding what others are aware of, practicing deception, and being aware of one's awareness are aspects of a higher level of consciousness that is more evident in birds and mammals than in other vertebrates. We cannot directly study the experience of consciousness in nonhuman animals, but we can gain clues to its existence by observing behavior and understanding the neural substrates that enable thoughtful, prospective actions. The varieties of behaviors we have reviewed clearly suggest a high level of consciousness in corvids.

Many of the fantastic behaviors we have attributed to corvids may depend on neural loops between the thalamus and forebrain that make reconsideration possible. But many other behaviors are certainly much less complex. Vocal imitation may require the ability to reconsider and update what a crow says with what it hears. This may only be possible because of the anterior forebrain circuit that loops between the forebrain and thalamus of a songbird. Screaming in the grasp of a predator is much simpler, relying perhaps entirely on direct stimulation of the midbrain. Scolding and mobbing a predator likely requires thought—in some situations it is withheld—but the actual vocalization is innate, not learned. Even linking innate scolding to a new danger, like the face of a person who wronged a crow, is a simple learning mechanism shared by all animals. Distinguishing among similar faces, associating biographical information with a face, and learning about such associations secondhand—all things that crows do—are more complex. Crows and people seamlessly apply their complex cognitive abilities to enhance simple cognitive tasks as needed.

The intriguing behavior of corvids tells us something about their mental faculties. Over time, comparative neuroscience will surely expand and modify much of what we have presented, but our general comprehension of the integrated way in which many regions of the bird's brain contribute to its decision-making ability will stand up to future discoveries. Understanding the mentality of other animals is in its infancy. And as with all our views on nature, we can assemble it only from our human perspective. We may never know what crows think about, but by understanding more about the anatomy, chemistry, and physics of their brains, we are learning something about how they may think. We may not, in this way, truly understand the mind of the crow, but we can begin to understand the brain of the crow. The glimpse we have revealed suggests that crows possess a brain capable of complex thought, which is consistent with an advanced state of conscious awareness. These animals, which we often take for granted and aggressively combat, really are thinking and reasoning in ways that are more similar to our own than many would care to admit.

Corvids assume characteristics that were once ascribed only to humans, including self-recognition, insight, revenge, tool use, mental time travel, deceit, murder, language, play, calculated risk taking, social learning, and traditions. We are different, but by degree. We both excel in large part because of our brains, our greatest gifts. To the crow in our world, where other beings are increasingly marginalized, their brains may also be a blessed curse. Sometimes when they get the best of us, we hold grudges, too, and thoughtlessly kill crows.

As crows affect our culture, so, too, do we affect their ecology, evolution, and culture. We are coevolving, each mutually shaping the other to varying degrees. Our cultural heritage with crows may even affect our mental processes, because the opinions we have formed about crows—their association with death, fidelity, creation, planning, thievery, and the like—are deeply embedded in our memory. The call of a crow in a film triggers an expectation in us that something ominous is about to happen. When we confront the loss of a loved one and expect crows to portend such loss, our subconscious sculpting of perception may fulfill our expectation. As an inescapable presence in our lives, crows have been codified to become part of our worldview. These animals inhabit our brains, and we are integrated into their brains as well. The caveman who caught crows, the researchers who have studied them, and the people who regularly feed them are stored as memories in many a crow amygdala or striatum. Our large brains apportion each other a bit of space. We are relevant and influential to each other. And our associations seem only to grow stronger and more diverse with time.

A good deal of what draws us to the corvids is their similarity to ourselves. They're not songsters, many are dressed in somber color, and often they are found in the most damaged of ecosystems, but their flexible innovations and social proclivity are intellectually stimulating. We never really know what we will learn as we count, observe, and challenge these birds in research settings. We are always intrigued by the things we see and hear corvids do. No day in the field is routine when your goal is to study a crow, raven, magpie, or jay.

But our interest in these magnificent creatures goes well beyond the scientific, artistic, and naturally historic. These birds cause us, and many others, to confront ourselves. Their behavior, so accessible to study, gets us thinking about the struggles and triumphs common to all life and what it takes to sustain it.

The charisma of corvids has made them sought after as pets and partners through history. We have heard from many people who grew up with a crow in the house. None would have traded the experience. Pet crows bond closely with people, expressing empathy, playing with them, and recognizing individuals. Most, like the one nurtured by Helen Stapleton's brothers in Pennsylvania, imitated sounds to their advantage. When the Stapleton crow wanted to hide, it retreated to a thick bush by the chicken coop and clucked, but when it wanted to get a rise from the pet dog, it gave a perfect rendition of the mom's voice to call the big shepherd over. Many a kid was followed to school by his pet crow. Some crows strolled along with their human companions, others flew, and a few rode on the bicycle handlebars. And after a daily routine of far-flung mischief, that crow would wait patiently for the school day to end so it could accompany the child back home.

Ken Botwright, now eighty-three, grew up in a small house with a police dog, a cat, a rabbit and—in the spring of 1943—Squawky the crow. Ken's brother procured Squawky from a nest high in a poplar tree near his Cochrane, Ontario, home. As he grew, Squawky bonded with Ken's mother, Sarah, following and often undoing her every attempt to keep the house in order. Squawky's curiosity nearly got the better of him one day as he chased the tail of a shirt heading into the clothes washer's ringer rollers and got a good pinch. Secluded for a couple of days in a dark closet, he recovered.

As with other crows we've met, Squawky soon learned his name.

"*Hello, Squawky,*" he would say to each family member. Ken's father, Reg, walked to work at the railroad every day with Squawky on his shoulder. Having a mind of his own, the crow never obeyed Reg's daily command to go home; instead he winged it from the freight office to the school where he'd join Ken and his brother for morning recess, and then go home. Squawky teased the cat, cached .22 bullets in beds and sugar bowls, and pestered the neighbors. When the butterscotch and white tabby slept on the couch, Squawky would tweak the feline's tail and then laugh, perfectly imitating Sarah's chuckle. A BB from a neighbor kid blinded him in one eye, a disability that Squawky adjusted to, mostly. He was killed a short year after the Botwrights obtained him. As he raced the family canine to a crust of bread, the big dog blindsided him. The family deeply mourned Squawky's death; Reg stayed home from work, Sarah let the housework go, and the boys skipped school to bury their favorite pet in the backyard.

Pet crows are sometimes too curious. Squawky was nearly crushed to death by his keeper's washing machine.

One year with a pet crow was etched deeply into the aging mind of Ken Botwright. He told his stories as if he were still holding the engaging bird. But the memory of this crow was stronger still in his mother's mind. As Sarah fought death thirty-three years after Squawky died, a mob of crows scolded outside the window of the North Bay, Ontario, hospital. The last words Ken remembers his mom saying were in response to the nearby corvid din: *"Oh, that Squawky."*

The crow's ability to bond with people has allowed corvids to hold prominent roles in our homes, legends, fables, and religions. And yet today keeping a pet crow—at least in North America—is illegal. This restriction comes from the Migratory Bird Treaty Act, an agreement between Canada, the United States, and Mexico that aims to protect 836 bird species from the destructive hand of humanity. But there is a fickle side to this important piece of conservation legislation. Some species, such as the American crow and many game birds, while protected, can be hunted under regulations established by individual states. In most states, crows can also be shot any time they are in the act of depredating crops. Falconers who train wild or captive-raised raptors to hunt birds, rabbits, and other game can legally possess other species protected by the Act, such as hawks and eagles. So, if crows can be killed and hawks can be kept, why can't an interested person legally possess a live crow? Should laws be changed to allow pet crows? We think they should.

We believe that, with proper regulation, a system fostering crows as pets could be devised. The North American Falconers Association provides a great model. A person interested in possessing a hawk must be knowledgeable and licensed. A two-year internship is required to begin the training process. Hawk-housing facilities must be inspected and approved. Only after seven years of close supervision can a falconer act with full independence. If we want to advocate for crow ownership, the first step is the development of a Crow Association that would compile reference materials for the safe and ethical adoption of crows, suggest requirements and steps in the certification process, and then lobby to change federal and state laws to allow regulated crow ownership.

As we ponder a future with pet crows, deep ethical questions must also be answered. Is it moral to enslave another sentient being for our pleasure? As a society, we have answered in the affirmative with respect to dogs, cats, monkeys, parrots (although a strong reaction against parrot ownership is growing), and more. In some cases, it may not even be simply for our pleasure that a crow is kept. Local wildlife rehabilitation centers are often overrun with abandoned and injured crows; many are euthanized. Some of this unfortunate surplus could certainly enliven an interested family. Wouldn't a life with humans be better than death? And with rehabilitation, a young foster crow could be released back into the wild after a few months of care. In our experience most young corvids are quick to sever ties with our species and rejoin their feathered brethren in the autumn when dispersal is natural. But even a short amount of time spent with a pet crow can profoundly stimulate and expand upon a love and understanding of nature in a young child or seasoned adult.

Pet crows are not for everyone. They get into no end of trouble—they will wake your neighbors, damage your and others' property, and, to varying degrees, rule your life. A friend's kitchen was completely remodeled by his crew of ravens. Such birds demand attention, nearly constantly. Raising a nestling crow requires sun-up to sun-down feeding, several times per hour. They are not potty trained, and their mess can stain your best outfit and furniture. The possibility of disease may worry some, but common ailments like pox are not transferable to humans, and recent ones such as West Nile virus quickly kill infected birds. If these drawbacks do not thwart your interest, you will find crows to be wonderfully curious, entertaining, and loving animals. Even those ravens that destroyed the kitchen also learned to turn the house lights off when they were ready to sleep.

We don't expect crows to be allowed as pets anytime soon, but this does not stop us from enjoying their antics in the wild. Few wild animals actually knock on your window or ring your doorbell for food. There may be no others that peer into windows, looking for those who have caused them harm. Leaving gifts and assembling dogs by imitating the master's voice are surely the provenance only of crows. Crows do not shy away from testing and using people to get what they need. They challenge us to say no. They invite us to continue our long natural legacy as a member of the biological community, not as one with dominion over it. Many people, including us authors, have accepted the invitation.

Join us in discovering remarkable things by watching crows. All you need, along with a pair of binoculars if you wish, is some patience and the employment of your senses. Write down what you see as soon as possible. Describe it completely so that others can learn from you. Then enjoy speculating and interpreting. Mixing interpretation with observation early clouds your discovery and makes it less useful to others, such as us. What you see is obvious, but look more closely; there may be more there than you realize. Listen; there are sounds that your crow subject is attending to that you need to describe along with its behavior. There may be other crows near and far from your subject that affect its behavior. Pay close attention to all the surroundings—biological, cultural, and physical.

You are likely to find corvids in practically any environment in which humans operate. Some in particular are easily entered and readily expressive of crow activity—behavior that is entertaining, instructive, and in the long run provocative as it makes us think about the remarkable possibilities of avian intelligence. Take the local park, for instance. There crows may seize on an opening to grab a sandwich from your table and cache it out of sight of the humans, squirrels, gulls, geese, and other crows intent on recovering a good meal.

Crows also seem to enjoy mind games with squirrels. In many parks squirrels pick up what they can from what's left by humans. Crows are quick to exploit the skills of the mammal that can plunge into the garbage. Often an attentive crow will wait for the squirrel to gain access to the food and then chase and harass it until the squirrel drops its prize. It is as if the squirrel is the latest crow tool.

On your next trip to the store or restaurant take some extra time to enjoy the various crow gatherings. Watch a crow size up the shoppers as they come and go. Is the human hostile, friendly, a good source of a handout, or indifferent? The crow's body language around people in these busy locations suggests to us its ability to read our intentions almost instantly. Of course the birds lean forward and approach people who offer something to eat, but they seem more tentative than the sparrows, pigeons, and gulls. The crows are cautious, shifting their weight back toward the escape route if it's necessary. As you revisit areas frequented by crows, you may begin to notice territorial and communal activities, the fission and fusion of crow life. In the shopping center the crows probably have apportioned off sections of real estate that specific birds claim for their own, not unlike the vendors who have shops there. Notice the postures and calls that defend crow turf. Of course all of this becomes moot when the birds hit the garbage at the back of the facilities and everyone dives in. Here the cost of defense is too great, so all share the rewards. We have always been impressed by the skills these birds use to get a tasty morsel from the depths of the dumpster.

Crows also flock to shopping and parking places as grand rallying points. In the fall and winter, just as dawn is breaking, crows rally and assemble in the open expanses of asphalt near their nocturnal roosts. Should you have such a gathering place, get up early and pull up a chair. You'll see crows greeting one another, seeming to form clusters of associates and probably planning the day's activities.

Crows are also interesting to watch in more natural areas. Where we live, crows and ravens use nearly two thousand miles of inland shoreline as a feeding ground. They dig clams at low tide and use the roadways and traffic to crush the mollusks. The birds dare to take from others what they need and will team up against the larger animals. Gulls, oystercatchers, eagles, otters, and seals are all able to catch and open up what the crows cannot, so crows apply their brains against the others' brawn with amazing results. As one appears to walk with indifference toward the feeding animal, another crow pulls the feeder's tail. The feeder drops its meal, which the pair of crows seizes and shares.

Sharing your own yard with a family of crows is nonstop excitement. If you discover a nest, you can observe all sorts of cooperative enterprise that corvids employ. There may be helpers, offspring from previous years raised by the pair of breeding birds, that contribute to the raising of their sibs and defense of the family from hawks, owls, raccoons, or cats. Their determination to protect and sustain their familial investment is an inspiration to behold. Has the crow figured out how to get access to the sunflower seeds and suet in the bird feeder, or does it hunt the other birds, small mammals, snakes, frogs, and insects that also live in your yard? The lives of predator and prey intertwine, and you will see strategy and counterstrategy shape these primal interactions.

In our neighborhood we have seen the reemergence of small raptors that rely on crows, not for food, but for shelter. The merlin is a small falcon that occasionally tolerates human settlement, but it is competitive and an anciently hostile associate of crows and jays. Nevertheless a remarkable tie binds the species. More often than not, merlins, who do not build nests, will use a crow's as a comfortable platform to hatch their eggs and raise their young. We've watched crows with nests two trees removed from the merlin family sit contentedly with their young on a telephone line as the parent falcons flew back and forth over their heads delivering food to their own clamorous brood.

Perhaps no other application of human technology has favored the crow and raven's expansion of population more than the interstate highway system. As you drive, look at the cornucopia of rendered food served up on an easily accessible plate of asphalt. You'll see crows and ravens emboldened to follow the hurrying traffic. And you can find new nesting strategies evolving along the roadways. Cell towers, power-line pylons, billboards, and outbuildings now support well-stocked nests. Even a lack of traditional nesting material doesn't slow down the innovative crows, which use everything from barbed wire to bailing twine to make nests (and short out power supplies on occasion) and line them with the hair they plucked from the roadkills or from the live cattle that wander over the plains to supply the appetites of the beef eaters.

Today, from these innovations and increases in many corvid populations, we see how our species has changed Earth. We also see how these animals can change people. Out of the blue we get letters and calls with stories: Caroline feels a deep connection to corvids; Denise

Not unlike pigeons, crows often befriend persons who consistently feed them.

is considering changing jobs because her business plans to move to a new office and she'll lose her view into an active raven nest; Levi has just released a new disc of music, and Jeanne a new set of cartoons, motivated by the comings and goings of their local crows. The crows in our lives continue to shape us as a species.

As we have explored its brain and emotional behavior, we have come to see the true gifts of the crow. Evolution has endowed the species with a big brain that empowers it to readily exploit and adapt to whatever we throw its way. This body of curious and intelligent beings has given us an invitation to develop a fresh relationship with nature, one that is no longer "multiply and subdue," but rather contemplate and learn. Contemporary and future generations will continue to enjoy these gifts and reap new ones.

Our realization that the nature we often take for granted includes animals that think and dream, fight and play, reason and take risks, emote and intuit may be the greatest modern gift of the crow. Companions since our birth as a species, crows, ravens, and their kin cause us to briefly step away from our current technological, urban lifestyles and to explore, understand, and appreciate nature. In so doing, they will continue to stir our souls and expand our minds.

Acknowledgments

OUR INVESTIGATIONS WOULD NOT have been possible without the contributions of many. We are especially thankful to the people who told us so much about their interactions with corvids, especially Susanne, Inda, and Abbe Drakborg, Gary Clark, Suzanne Wyman, Julianne and Marco Restani, Juliana and Carole Ann Coffey, Glenn Rimbey, Ken Botwright, Valerie and Tom Allmendinger, Katie Roloson, Jack Ingram, Kevin Smith, Lijana Holmes, William Pochmerski, Kay Schaffer, Russ and Judith Balda, Chuck Wischman, Cathleen Handlin, Rita Ross, Darla Dehlin, Mary and Jim Ronback, Lydia and Stuart Janik, Jack Fellman, Porter Evans, Jeanette Griver, Helen Stapleton, Larry Makinson, James Anderson, Shelly and Jim Leonard, Phyllis Alverdes, Carl Marti, Prisca Cushman, Jingle Ruppert, Gene Carter, Ted Hayes, Nancy Kool, Bill Romaniuk, Katrina Anderson, Randy Gerber, Donna Barr, Nick Mooney, Jene Jernigan, Barbara Arnold, Janet Brandt, Evon Zerbetz, Mary Palm, Richard Borgen, Carol Strickland, Barbara Brozyna, Madeline Kornfield, Jim Schumacher, Crow Swimsaway, Mary Coggins, Robert Mooney, Dana Casey, Emilie and George Rankin, Deborah Raymond, David Caley, Julie Lawell, Kenneth Geirsbach, Padmini Pooleri, Josh Mong, Kent Bush, Cal Aylmer, Denise Storm, Levi Fuller, Jeanne Shepard, Gretchen Copland, Cathi Winings, Katharine Stieg, Regina Rochefort, David Williams, Mark Young, Bill McCuaig, Cori Conner, Mason Reid, Johan Peleman, Evan Evans, Dorothea Forrest, Elwin

Wright, Noni Pope, Lois Magnusson, Gail Olson, David and Cindi Sonntag, Sandy Harbanuk, Jay Dennett, Bob Armstrong, Beverly McCann, Daryl Clark, Lela Brown, Mike Przystal, Richard Buchholz, Renee Delight-Latorre, Cathy Fields, Tamara Pawlak, Joe Roberts, Russ Furbush, Andy Edgar, Carol King, Andree Dubrevil, Shannon Armstrong, Kip Schwarzmiller, Loren Haury, Rachel Landry, Diana Constable, Cindy Zaring, Rita Loehr, Somesh Dixit, Patrick Hemphill, Gus van Vliet, and Cone Johnson. Our ongoing conversations with scholars, artists, and naturalists including Bert Bender, Ivan Doig, Tom Quinn, Les Perhaes, Larry Schoonmaker, Pat Lewis, David Musell, Marc Miller, Al Harmata, Eliot Brenowitz, David Perkel, Rae Kurosawa, Barbara Clucas, Russ Balda, Bernd Heinrich, Al Kamil, Boria Sax, Matt Betts, Barbara King, Corina Logan, Donna Cross, Robert Miyaoka, Greg Garwin, Barbara Lewellen, Robert Peck, Fred Lohrer, Catherine Connors, Bill Webb, Arne Zaslove, Jim Kenagy, Donna Darm, Jeff Williams, Ken Raedeke, and Bill Harrower provided encouragement and refined our thinking and creative approaches. Colleen Marzluff, Lee Rolfe, Michael Marzluff, David Perkel, Marco Restani, Tom Rivest, Kristy Gould, Kezia Toth, and Marg Leehane read our early manuscript and greatly clarified, sharpened, and corrected our thoughts. Greg Krogstad brought clarity to our technical diagrams with bold and original text. Mary Ann Jeffreys and Donna Loffredo provided timely and much appreciated editorial and technical assistance from Free Press. Our agent, Trena Keating, and editor, Leslie Meredith, were instrumental in shaping our ideas and approaches to telling a large audience about the birds that so engage us. And finally, but by no means least, we wish to thank our wives and families. Colleen Marzluff and Lee Rolfe Angell have given generously of their time, energy, and expertise to clarify our thinking and move our efforts forward. Our daughters Danika and Zoe Marzluff, and Gilia, Bryony, Gavia, and Larka Angell have provided timely inspiration, assistance, and patience to advance our corvid obsessions.

Appendix

Supplemental Images

IN THE IMAGES that follow, you can see our rendition of a neuron, the central nervous system, and the brain, in which we highlight important anatomical structures and processes. Each image and its caption is intended to provide the detail we summarize in the text. We begin with a neuron and the central nervous system generally, then move to the brain, illustrating its basic construction, sensory inputs, neural loops between forebrain and thalamus, specific sensory input for sound and loops for speech, complex connections among neurons from memory, thought, and emotional centers, and regions important to the social brain. Our diagrams represent the three-dimensional anatomical features in only two-dimensions, which, while generally accurate, are therefore not precise.

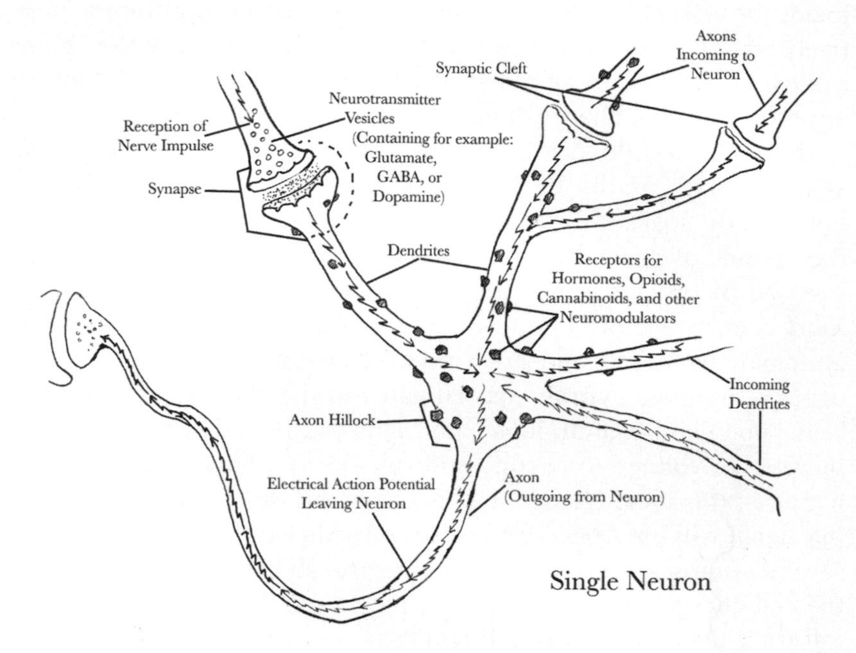

The nuts and bolts of brainpower.

THE COMPLEX BEHAVIOR of animals derives from relatively simple behavior of their neurons. Neurons have an enlarged cell body that houses the nucleus, short dendrites that accept connections from other neurons, and long axons that carry electrical information on to other cells. Neurons connect to each other and to muscle cells at junctions between dendrites and axons called synapses. Each neuron can have hundreds or thousands of synapses, and synapses can be created or lost with experience; maintaining such connections is part of the memory process. Severing or stifling connections is the cellular mechanism of forgetting and ignoring. When old connections are no longer useful, they are retired or eliminated and replaced with new connections that serve current needs.

Neurons convert into electricity the colors and shades of light; the mix of sweet, sour, spicy, and pungent chemicals tasted or smelled; and the pressure waves from sound or touch that we or a bird senses. Neurons do this by changing their electrical charge. At rest, without stimulation, a neuron is electrified because its outer membrane maintains a greater positive charge outside relative to inside the cell. This resting potential is maintained by pumping positively charged sodium ions out of the cell. If we measure the charge across the membrane of a resting cell, it would be about -70mv, or around 1/20th the charge of an AA battery.

Nerve cells influence each other so electrical signals are propagated along circuits of connected neurons. Electrical energy is not directly sparked from cell to cell. Rather, it is transmitted by the actions of special chemicals called neurotransmitters that are released by other neurons into the synapse. One of the most common neurotransmitters in vertebrate brains is glutamate. When glutamate from one neuron's axon binds with receptors on the dendrites of another neuron, special gates in the receiving cell's membrane open and sodium ions enter. This causes a slight and positive increase in voltage to occur, from say -75 to -65mv. If nothing else happens, this local change in charge will be short-lived; the incoming signal will have been filtered from further influence. However, if several synapses—typically five to twenty—all bind glutamate so that the cell charge at many dendrites increases, this change begins to influence the entire neuron. When the summed influence of the dendrites causes the membrane charge at the axon hillock to reach about -60mv, then charge-sensitive gates open and sodium really pours in, causing a rapid and short-lived spike of charge up to about +50mv. A speeding spike of electricity—the fundamental unit of information in the nervous system—runs down the axon in about ten milliseconds, causing a wave of gate opening that perpetuates the spike and eventually allows positively charged calcium ions to enter the terminal end of the nerve cell. It is this calcium that triggers the release of neurotransmitters that flood into the synapse and, if sufficient, stimulates an electrical spike in the next cell in the information circuit.

Synapses figure importantly into our understanding of how the nervous system shapes behavior. The reception of various neurotransmitters—there are hundreds—determines whether a signal is increased, decreased, or filtered out by the nervous system. As mentioned above, glutamate is a common, excitatory neurotransmitter. As it binds to the glutamate-specific protein receptors on a neuron, it functions to increase the frequency of spikes, or firing rate, of the neuron. Glutamate helps move and preserve important messages in the brain. Gamma-aminobutyric acid (GABA) is another common brain neurotransmitter that does just the opposite; firing rates are slowed or inhibited when dendrites bind GABA. Acetylcholine is an excitatory neurotransmitter that links neurons to muscle cells, controlling the final behavioral response to the brain's electrical messages. Dopamine and serotonin are neurotransmitters with many functions. They influence neurons directly, affecting mood and attention, and indirectly by increasing or decreasing the response of neurons to other neurotransmitters. In crows and people, both are critical. Low dopamine, whether from Parkinson's disease or other afflictions, reduces our motivation. Serotonin balance is critical to brain-stem regulation of our sleep and wake cycles. Too little serotonin results in depression and difficulty sleeping.

Neuron cell membranes have many specific receptors, or binding sites, for hormones and other chemicals like opioids, cannabinoids, and dopamine to affect the ease with which electrical signals pass from one neuron to the next. This synaptic plasticity is a critical feature of learning, or changing the connections between streams of information within the brain.

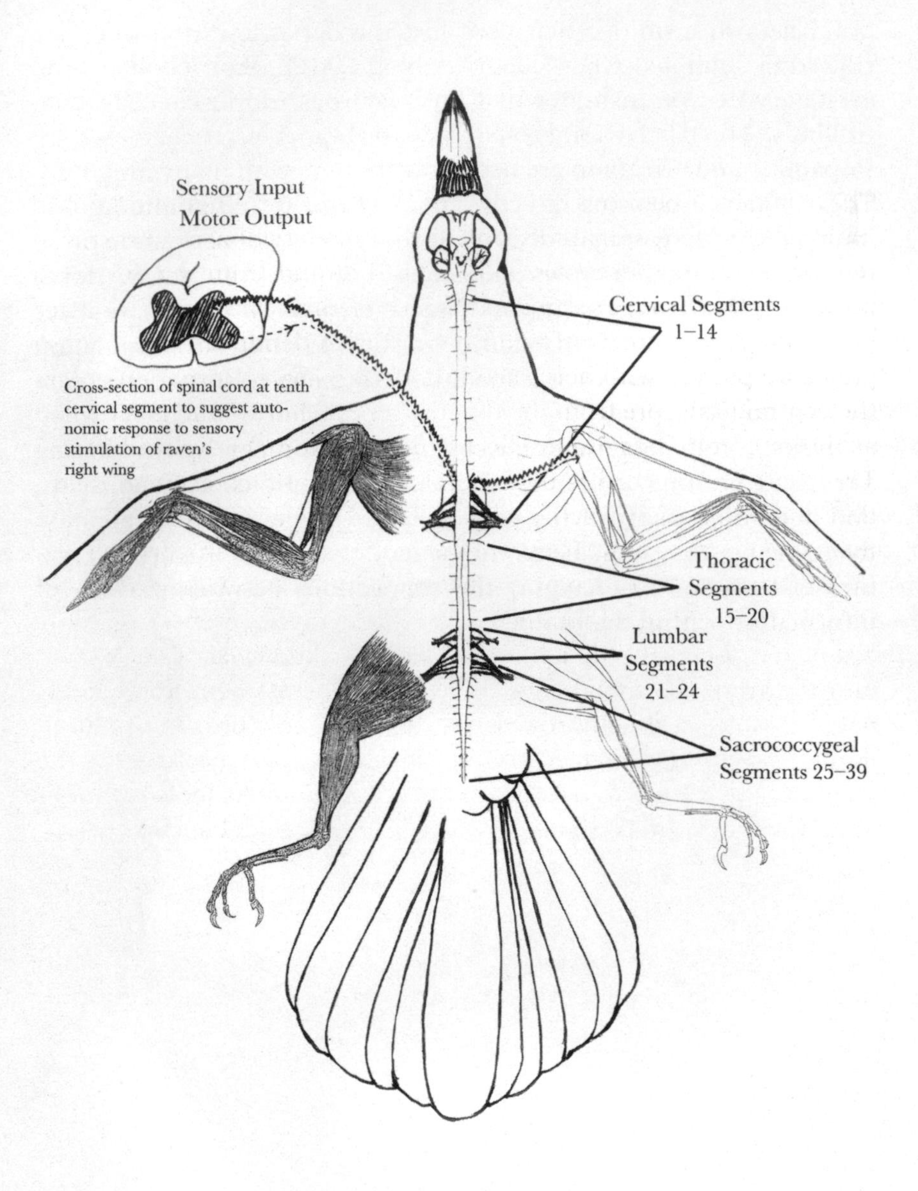
Sensory Input
Motor Output
Cross-section of spinal cord at tenth cervical segment to suggest autonomic response to sensory stimulation of raven's right wing
Cervical Segments 1–14
Thoracic Segments 15–20
Lumbar Segments 21–24
Sacrococcygeal Segments 25–39

Appendix

THE CENTRAL NERVOUS SYSTEM of all vertebrates, including birds, consists of the spinal cord and the brain. The spinal cord assesses, reacts, and conveys sensory information to and from major nerves and the brain. A cross section of the spinal cord reveals a white outer layer composed mostly of neuron axons and dendrites and a dense gray core packed with nerve cell bodies. Sensory information enters the central gray core from the top, the "dorsal horn," and is returned as motor commands to muscles from the bottom, or "ventral horn." The bird's spinal cord is adapted to control and coordinate flight; therefore we see enlarged and merged spinal nerves that innervate the wings and the legs. These nerves carry information about movement to the spinal cord, which may immediately react and command muscles to respond, or forward information to the cerebellum in the brain along specialized, high-speed conduits known as "Clarke's columns" within the spinal cord for reasoned assessment and coordinated action. In the lower reaches of the spinal cord, information flows in from the skin, feet, and wings. In the upper neck region the spinal cord expands to become the brain stem and is fed with information about sounds, sights, and smells from cranial nerves.

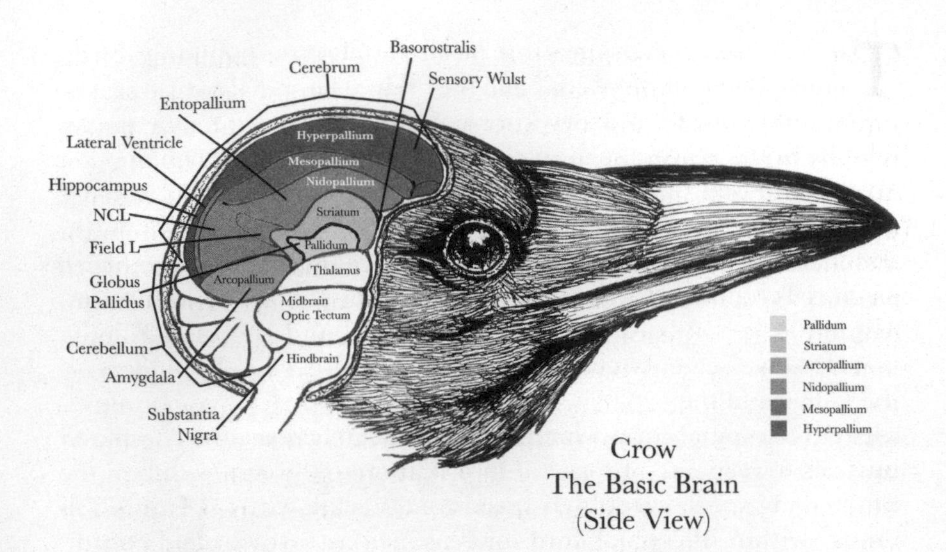

Crow
The Basic Brain
(Side View)

THE BRAIN OF A BIRD, like our own, comprises a hindbrain, midbrain, cerebellum, thalamus, and forebrain, or cerebrum. The forebrain of birds and mammals consists of striatal and pallial regions derived from our earliest vertebrate ancestors. The interior of a bird's forebrain includes the pallidum and striatum, where emotions, costs, and benefits shape behavior. The base of the forebrain includes the arcopallium, which ushers commands to muscles to produce behavior. The majority of the forebrain consists of the nidopallium, which integrates and synthesizes reaction to sensory inputs much like an executive would determine what actions to take. Atop the nidopallium lies the mesopallium, and finally the hyperpallium, which includes a forward portion known as the Wulst to process sensory and motor information, and a central portion, or hippocampus, which maintains contextual memories. Specialized areas within the brain consider distinct sensory information including vision (optic tectum, entopallium, and hyperpallium), touch, taste, smell (basorostralis), and sound (field L).

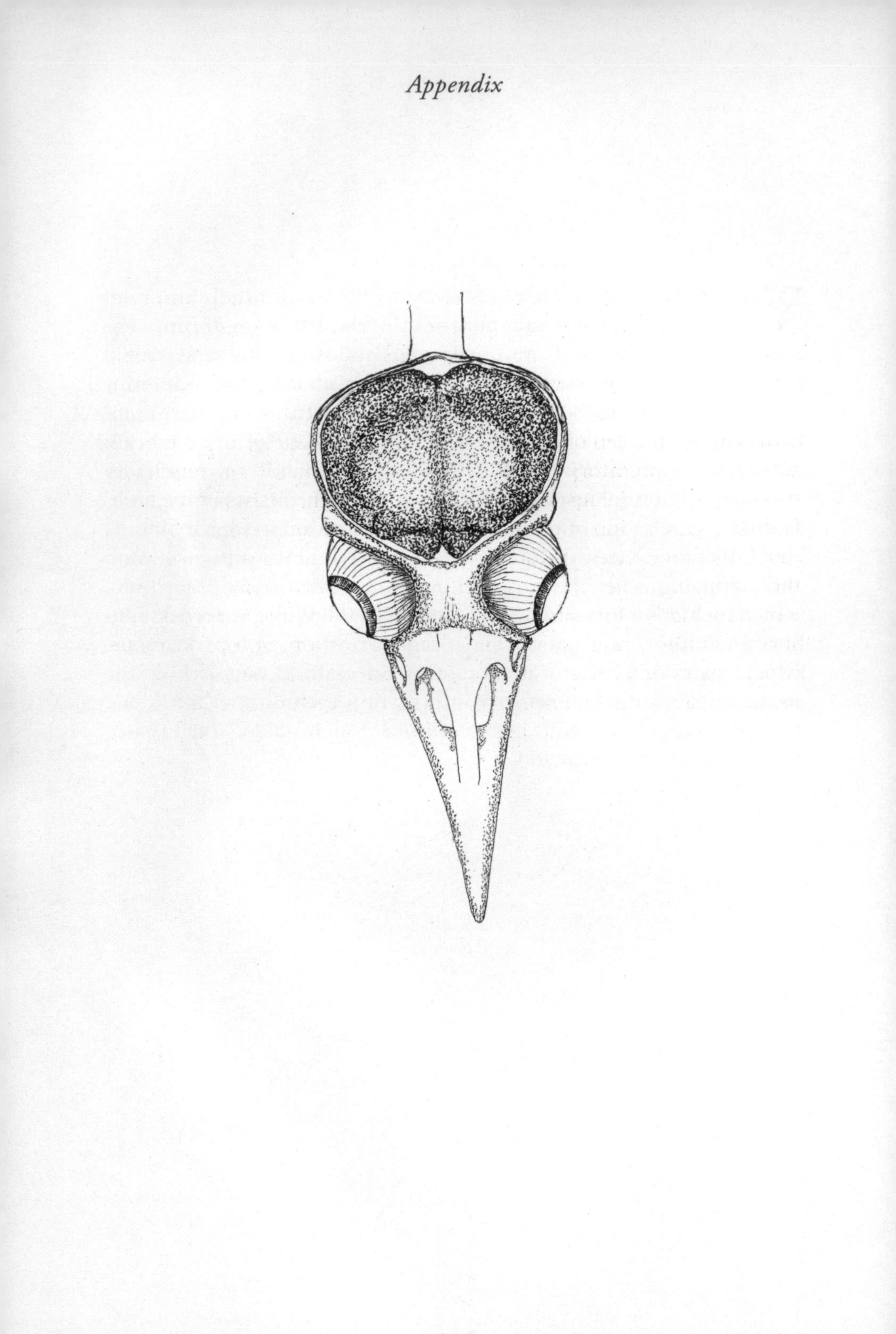

VIEWED FROM ABOVE, the two hemispheres of a bird's cerebrum are evident. Unlike in mammals, these hemispheres are not strongly connected in birds (birds have no corpus callosum), although many neural circuits cross between them. As a result, the two sides of a bird's brain work more independently than do those of a mammal. Birds can even sleep one side of their brains at a time, a handy trick during long migratory flights. Some tasks in birds and mammals are done predominantly by either the left or right brain hemisphere. The binocular vision of birds when looking forward, which may help identify human faces, distinguish grains of food from those of sand, and enable the fine motor skills needed to fashion serviceable tools, is processed mainly from the right eye to the left brain hemisphere. In contrast, the monocular view of the side of a bird's eye, which is important to detection of predators, is primarily done with the left eye and right brain hemisphere.

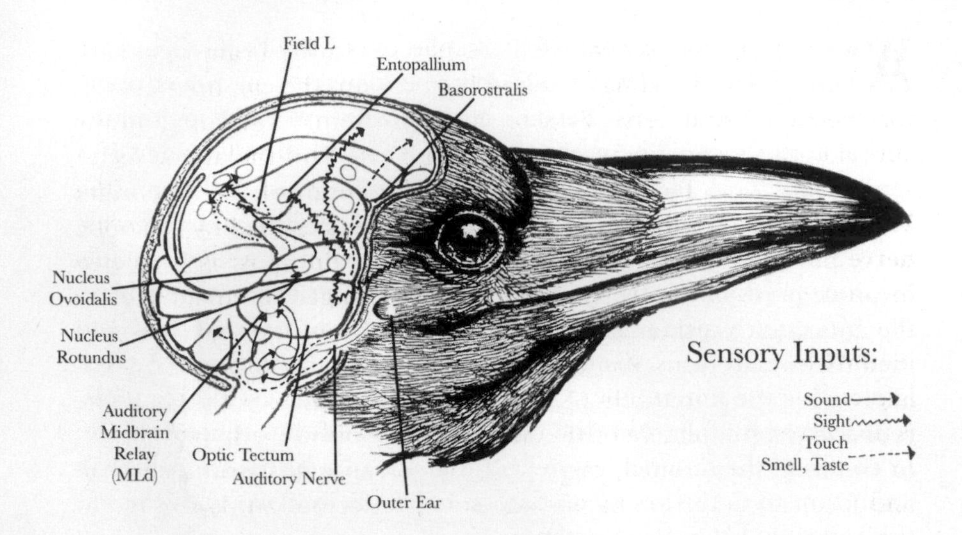
Field L
Entopallium
Basorostralis
Nucleus
Ovoidalis
Nucleus
Rotundus
Auditory
Midbrain
Relay
(MLd)
Optic Tectum
Auditory Nerve
Outer Ear
Sensory Inputs:
Sound
Sight
Touch
Smell, Taste

A WEALTH OF SENSORY information flows into the brain of a bird. The electrical coding of sound rushes from the ear, up the auditory nerve (cranial nerve VIII) to the brain stem, midbrain, and the superior olive ganglia within the thalamus. The thalamus relays sensed sound to Field L within the nidopallium of the forebrain. Visual information from the retina of the eye is carried by the optic nerve but then takes two routes to the forebrain: most is projected in an organized way (literally, spatially arranged as an image) to the optic tectum, then to the nucleus rotundus of the thalamus and then the entopallium. Some visual information, in crows and ravens, likely their forward-facing, binocular view, goes directly from the retina to dorsal ganglia of the thalamus and on to the hyperpallium of the forebrain. Smell, taste, and touch sensation from the beak and face, and even perhaps some sound information, is carried to the brain stem on the Vth cranial nerve known as the trigeminal nerve. This multimodal nerve sends some sensory information about the beak directly to the basorostralis of the nidopallium, which may be quickly processed and used in combination with the trigeminal information that is simultaneously sent to the cerebellum to command the many neck and head muscles that enable coordinated feeding action and reflexive head movements, such as the instantaneous orientation of an owl's head to the sound of a rustling mouse. The trigeminal nerve also sends information to the thalamus that is relayed to the Wulst region of the hyperpallium. Routes of sensory information in and out of the brain are much more complex and less discrete than we illustrate.

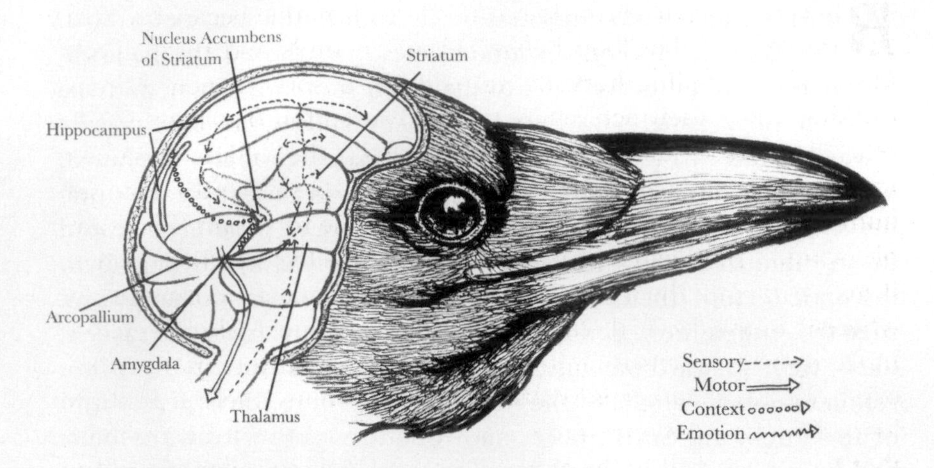
Nucleus Accumbens
of Striatum
Striatum
Hippocampus
Arcopallium
Amygdala
Thalamus
Sensory
Motor
Context
Emotion

Appendix

BIRDS ARE ABLE to reconsider sensory information because of neural circuits that loop between the thalamus and the pallium. Mammals, including humans, use the same sorts of neural loops to think about their actions. Sensory information that is originally relayed from the thalamus to multiple areas in the pallium (as shown in the previous illustration, including Field L, basorostralis, entopallium, and Wulst) is further routed through the forebrain (for example, sound from Field L goes to the HVC of the hyperpallium) where it may stimulate the arcopallium to relay commands to muscles, or where it may be joined with emotional and contextual information in the striatum. Sensory information shaped by emotion and memory from the striatum is passed to the thalamus where it is again looped back to the sensory areas that produced them. Even signals that have been sent to the arcopallium to drive muscular action can be reconsidered and further shaped. As these motor commands are issued, they stimulate the thalamus with essentially what are copies of the commands, or "corollary discharges." The thalamus restimulates sensory areas with information from the striatum and from the arcopallium to provide a bird's brain with an expectation of what its body will do and the significance of the action. This allows expectation and realization to be compared mentally and behavior adjusted to its outcome.

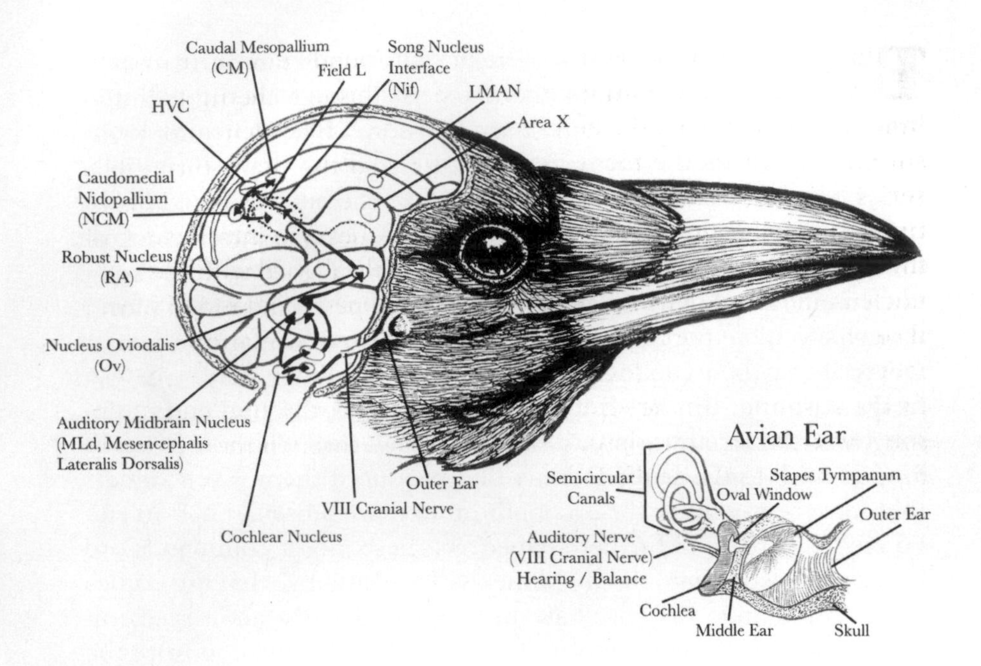
Caudal Mesopallium
(CM)
Field L
Song Nucleus
Interface
(Nif)
LMAN
Area X
HVC
Caudomedial
Nidopallium
(NCM)
Robust Nucleus
(RA)
Nucleus Oviodalis
(Ov)
Auditory Midbrain Nucleus
(MLd, Mesencephalis
Lateralis Dorsalis)
Outer Ear
VIII Cranial Nerve
Cochlear Nucleus
Avian Ear
Semicircular
Canals
Stapes Tympanum
Oval Window
Outer Ear
Auditory Nerve
(VIII Cranial Nerve)
Hearing / Balance
Cochlea
Middle Ear
Skull

THE CROW'S AUDITORY PATHWAY begins as sounds tap on the outer ear, setting up vibrations that proceed through the middle and inner ear to stimulate the eighth cranial nerve. Electricity, not pressure, now carries the message of sound into the brain through a series of brain-stem and midbrain relays to the nucleus ovoidalis in the thalamus. From here neurons carry the message into Field L of the nidopallium, which is a diffuse area closely aligned with the song nucleus interface (Nif), and two areas in the meso- and nidopallium thought to be important to memory storage and formation (caudal mesopallium and caudomedial nidopallium). From here, primarily through the Nif, the electrical code for a heard sound enters the song-learning and song-production networks through the hyperpallium region known as the HVC.

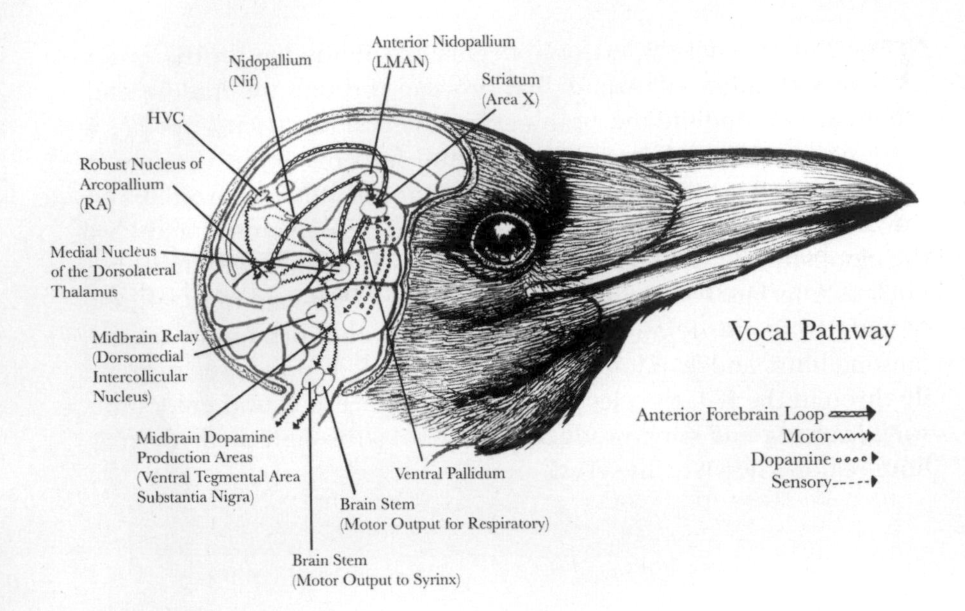

Anterior Nidopallium
(LMAN)
Nidopallium
(Nif)
Striatum
(Area X)
HVC
Robust Nucleus of
Arcopallium
(RA)
Medial Nucleus
of the Dorsolateral
Thalamus
Midbrain Relay
(Dorsomedial
Intercollicular
Nucleus)
Midbrain Dopamine
Production Areas
(Ventral Tegmental Area
Substantia Nigra)
Ventral Pallidum
Brain Stem
(Motor Output for Respiratory)
Brain Stem
(Motor Output to Syrinx)
Vocal Pathway
Anterior Forebrain Loop
Motor
Dopamine
Sensory

Appendix

A SIMPLIFIED, GENERALIZED diagram of the forebrain song-learning circuit of songbirds. Once a sound enters the HVC, it can proceed in two distinct directions: (1) to the arcopallium and on to produce a vocal response by stimulating the twelfth cranial nerve, which controls the syrinx and another set of nerves controlling the muscles needed to breathe; and (2) to the striatal region known as Area X. Here we begin the anterior vocal learning pathway, a type of loop between the thalamus and the forebrain that enables a songbird to reconsider and sculpt the electrical code for an utterance before it is sent to the arcopallium to produce a sound. Sounds sent from the HVC into the forebrain learning pathway are those that a bird will imitate. Some neurons in the HVC have mirrorlike properties—vocal stimulation of the neuron is directly translated into the electrical signal that would properly stimulate the syrinx to produce the sound. Sending mirrored signals into the vocal learning pathway helps seed the learning process with an accurate model. But that is just the beginning. In this loop, which is more complex than shown and of which our knowledge is expanding daily, neurons link Area X to the DLM relay in the thalamus, which returns the message to LMAN region of the nidopallium. As the code loops between LMAN, Area X, and DLM, midbrain regions may be stimulated to release dopamine into the striatum, including Area X. Here the arrival of dopamine may indicate the similarity of the current code to songs that are being heard and motivate the learner to adjust its song more precisely to that of its tutor. Dopamine also facilitates new and stronger synapses between neurons whose firing pattern will codify the bird's voice. At some point the current rendition of a learned sound moves from LMAN to the arcopallium and onward to the muscles around the syrinx, lungs, and air sacs to enable speech and song. There is at least one more way that an utterance can be reconsidered by the forebrain of a songbird. That is by corollary discharge from the arcopallium. In this process a copy of the neural command sent toward the syrinx and lungs reverberates to the DLM of the thalamus and eventually to Area X where it can be used as a benchmark of current voice to which new sounds coming from the HVC can be compared.

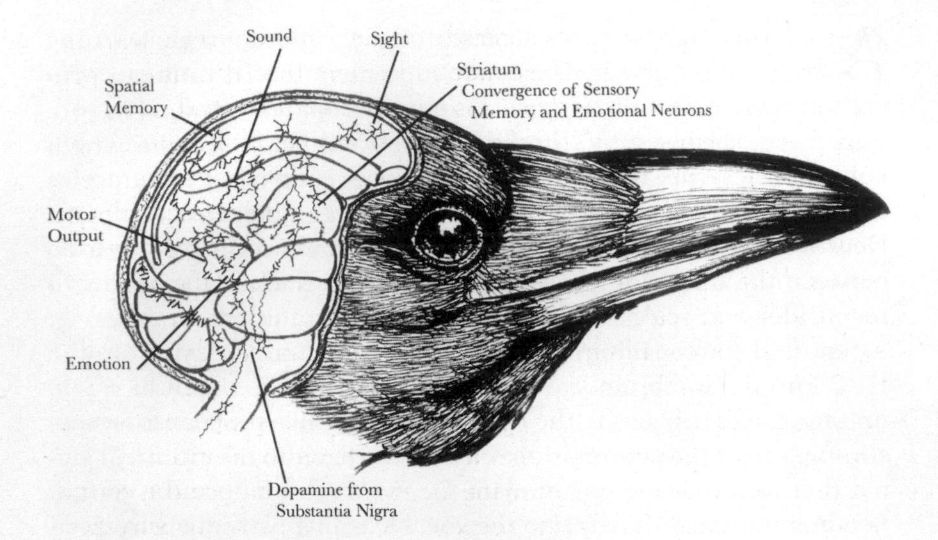
Sound
Sight
Striatum
Convergence of Sensory
Memory and Emotional Neurons
Spatial
Memory
Motor
Output
Emotion
Dopamine from
Substantia Nigra

Appendix

COMPLEX SYNAPSES BETWEEN neurons from different regions in the brain enable crows to associate aspects of the environment with consequence and inform these associations with context and emotion. These synapses are learned associations that guide subsequent behavior. Here we show a cartoon of how several neurons might synapse in the crow's striatum, for example in the nucleus accumbens. Neurons that respond to the sights and sounds of being captured connect to the striatum from integrative areas such as the NCL, primary sensory areas like Field L, or from integrated midbrain (optic tectum) responses. In the striatum these neurons could synapse with neurons responsive to context (hippocampus) and emotion (amygdala), as well as those from other forebrain regions. Synapse formation and the firing of neurons may be increased by hormones such as corticosterone and epinephrine from the adrenal gland response to stress, and from dopamine responses to environmental surprises. The neural circuit forged by synapses among neurons from different regions of the brain may stimulate the arcopallium or Wulst to usher motor commands to muscles that result in appropriate behaviors. For example, the trauma of capture links a negative experience with the sights and sounds of the net gun, food, and place into a neural circuit so that when these sights and sounds are again encountered, the firing of neurons in the network command the bird to flee or become increasingly wary.

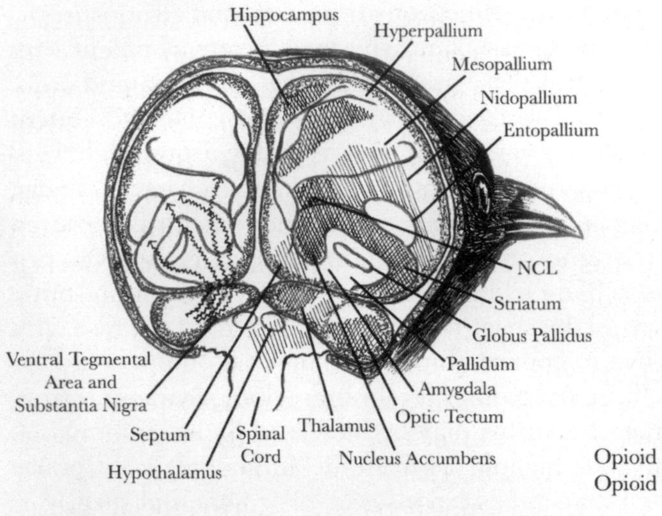

Opioid Receptors μ
Opioid Receptors κ
Both
Dopamine

Appendix

IMPORTANT REGIONS in a bird's social brain network are best seen from a cross section at the center of the bird's brain. This reveals the deep and more centrally located regions that are especially active in response to social situations such as bonding with partners, fearing enemies, and playing. Regions and places in these regions where opioid rewards for many social behaviors bind to neurons are indicated on the crow's right hemisphere (they occur equally on the left hemisphere). Opioids originate in the hypothalamus. On the left hemisphere, we show the origination of dopaminergic neurons in the midbrain coursing to the striatum and pallial areas where this chemical influences integrative synapses (this also occurs on the right hemisphere). Other chemicals are also critical to the activity in the social brain; stress hormones that bind to the brain stem and midbrain affect the subsequent release of serotonin and norepinephrine, both of which affect learning and activity during social interactions. Junctions among neurons from many places in the social brain network connect contextual memory stores from the hippocampus and emotional response from the amygdala with sensory information to affect neural activity in socially responsive areas such as the septum. Many social behaviors arise as the combined influence of all of these areas on neural commands to the muscles that control movement and vocalization.

Notes

References Made within the Preface

BATES AND BYRNE (2007) suggest that anecdotes can be important to the study of animal behavior especially if they are made by experts, recorded immediately after observation, stick to description rather than interpretation, and are accumulated among many observers. We agree with all except the need for reliable observations to be made by experts. We include many observations made by ordinary citizens because they frequently encounter crows and ravens and in our opinion provide accurate descriptions of rare behaviors in many cases. We tried to assure this in the anecdotes we present by talking with observers and especially by emphasizing accounts that were recorded at the time of observation and by multiple witnesses. McKelvey et al. (2008) also discuss the importance of proper consideration of anecdotes.

Our comprehension of the bird's complex brain is increasing daily, and much of what we discuss will doubtless be refined over the next few decades. If you are interested in tracking this progress, keep an eye and ear tuned toward the insightful writings of avian neuroscientists like David Perkel, Georg Striedter, Ann Butler, Andrew Iwaniuk, Luis Puelles, Anton Reiner, Onur Güntürkün, and James Goodson.

References Made within Chapter 1: Amazing Feats and Deep Connections

Research about Betty the New Caledonian crow is reported by Weir et al. (2002). In the online material to this article in the journal *Science* is a video of Betty solving the problem we describe.

The phrenological bust of the crow is from Streither (2005).

The metaphor of a corvid mental tool kit is from Emery and Clayton (2004a). Their ideas are more fully developed in Emery (2006), Emery and Clayton (2004b), and Emery et al. (2007).

The natural histories of the rare Mariana, Hawaiian, and Banggai crows are detailed in *The Handbook of Birds of the World,* Vol. 14 (corvid family is profiled by dos Anjos et al. (2009).

Our theory of cultural coevolution between people and corvids was developed generally in Marzluff and Angell (2005a) and formally in Marzluff and Angell (2005b).

The story of Dickens and Barnaby Rudge was compiled from Dickens (1900), Capuzzo (1993), and Velella (2009). In addition, we consulted with Robert Peck, Fellow of the Philadelphia Academy of Natural Sciences, about the Academy's preparation and presentation of Grip.

We quote from Edgar Allan Poe's poem *The Raven,* published in the *New York Evening Mirror* in 1845 and now in the public domain.

References Made within Chapter 2: Birdbrains Nevermore

Russ Balda's work on the spatial memory of nutcrackers includes Balda and Kamil (1992), Balda et al. (1996), and Kamil et al. (1999). His work on the mental abilities of the social pinyon jay includes Bednekoff and Balda (1996), Bond et al. (2003), and Pas-y-Mino et al. (2004).

The tendency of pinyon jays to carefully select mates based in part on their prior accomplishments is reported in Marzluff and Balda (1988a, b).

General introductions to the functioning and study of animal and human brains include the following books: Libet (2004), Striedter (2005), Linden (2007), and Allen (2009).

Bernd Heinrich's string-pulling experiments are reported in Heinrich (1995 and 2000), the deceptive caching studies are from Heinrich and Pepper (1998) and Bugnyar and Heinrich (2005), and the gaze-following study is from Bugnyar et al. (2004). Heinrich's general insights into corvids are discussed in his book *Mind of the Raven* (1999). Our work with Bernd on raven roosting and foraging strategies is detailed in Marzluff et al. (1996) and Marzluff and Marzluff (2011).

Nicky Clayton and Nathan Emery's research on scrub-jays includes Clayton and Dickinson (1998), Clayton et al. (2001a, b), and Emery et al. (2004). The general ability of birds to represent past events is discussed by Crystal (2010).

The ability of magpies to remember what, where, and when they cache foods was demonstrated by Zinkivskay et al. (2009), and their ability to imagine objects is reported by Pollok et al. (2000).

The basic working of neurons to transmit information and adjust their activity and connections as memories are formed is reviewed by Linden (2007), Allen (2009), and Kandel (2009). Some molecular bases of memory and learning are reviewed by Kotaleski and Blackwell (2010). The workings of place neurons in the hippocampus and as birds forage is reviewed by Sherry and Mitchell (2007) and Adams-Hunt and Jacobs (2007). The neurobiology of emotion is reviewed by LeDoux and Phelps (2008), and Wager et al. (2008).

The general physiology and morphology of the bird's nervous system and routes by which sights, sounds, smells, and tactile sensations reach the brain is from Whittow (2000).

Sensing of magnetic fields by birds is discussed by Heyers et al. (2010), Zapka et al. (2009), Ritz et al. (2004). General use of a variety of senses to migrate is reviewed by Wiltschko and Wiltschko (2009).

Striedter (2005) discusses and illustrates the comparative evolution of vertebrate brains. Butler (2008) provides the timetable and cladogram for vertebrate brain evolution that we discuss and illustrate in figure 10. Reiner et al. (2005) discuss the evolution of the bird forebrain in detail.

The evolution of mammal and bird brains, focusing on the embryonic homologies within the pallium and how and why these occur is discussed by Aboitiz (2011). Nomura et al. (2009) show how Reelin directs other genes to produce the layered mammalian and patchy avian forebrain. For example gene Brn2 is expressed in layers II, III, and IV of the mammalian neocortex and the nidopallium of the bird.

Ludwig Edinger's life is profiled by Glees (1952) and Stahnisch (2008). Konrad Lorenz was born in 1903, and Edinger died in 1918. Lorenz (1937) reports on the jackdaw response to bathing trunks and the raven response to held companions. He synthesizes his evolved views of instinct and learning in 1981.

The revolutionary story of modernizing our view of the bird's brain is told in Reiner et al. (2004) and the Avian Brain Nomenclature Consortium (2005). An overview and updates can be found online at www.avianbrain.org.

The role of the NCL and CDL as the bird brain's executive center is discussed by Hartmann and Güntürkün (1998), Güntürkün (2005), Rose and Colombo (2005), Butler and Cotterill (2006), and Kirsch et al. (2009). The ability of NCL neurons to reflect the amount of food and the delay to acquire it is from Kalenscher et al. (2005). Executive function in the mammalian prefrontal cortex is reviewed in Lefebvre et al. (2004), Striedter (2005), Linden (2007), and Allen (2009).

Dopamine is known to come from cells in the midbrain, specifically the substantia nigra and ventral tegmental area, and to influence synapses in the nucleus accumbens of the striatum and the NCL (Kubikova et al. 2009).

The hippocampus of birds and mammals is contrasted by Benhamou and Poucet (1996), Maguire et al. (1998), Bingman and Able (2002), Striedter (2005), Bingman and Gagliardo (2006), Nardi and Bingman (2007), and Allen (2009).

Sensory routes are from Whittow (2000), supplemented by Laverghetta and Shimizu (2003), Jarvis (2004), Avian Brain Nomenclature Consortium (2005), Mouritsen et al. (2005), Theunissen and Shaevitz (2006), Heyers et al. (2007), and Bauer et al. (2008).

Comparative brain size is shown for animals generally by Emery and Clayton (2004b), Rogers (2004), Striedter (2005), and Allen (2009). Variation in brain size among birds is discussed by Wyles et al. (1983), Lefebvre et al. (1997), Iwaniuk and Nelson (2003), Marzluff and Angell (2005a), Sol et al. (2005, 2007), and Lefebvre and Sol (2008). Special consideration of corvid brain size and the size of various regions in the corvid brain relative to other species can be found in Emery and Clayton (2004b), Marzluff and Angell (2005a), and Cnotka et al. (2008). Mehlhorn et al. (2010) discuss the nidopallium and mesopallium in the New Caledonian crow.

Energetic costs of the brain are discussed in Linden (2007), and its linkage to high-quality diet is reviewed by Striedter (2005) and Allen (2009).

The looping of information between the forebrain and thalamus is discussed generally by Farries (2001, 2004), Montagnese et al. (2003), Butler et al. (2005), Butler and Cotterill (2006), Llinás and Steriade (2006), Butler (2008), and Feenders et al. (2008).

Generation of new neurons in the bird's forebrain is discussed by Nottebohm (1984), Alvarez-Buylla and Nottebohm (1988), and Hornfeld et al. (2010). A comparative perspective on how the brains of birds and other species change with the season is provided by Tramontin and Brenowitz (2000). That corvid hippocampal regions reflect the demands of seed caching is from Healy and Krebs (1993), Krebs et al. (1995), Basil et al. (1996), and Patel et al. (1997). More recent investigation of how the hippocampus changes seasonally to reflect seed caching in a bird (chickadee) is provided by Sherry and Hoshooley (2009). Professor Kristy Gould, at Luther College, has documented the seasonal growth of the jay hippocampus as the birds focus on caching.

Sleep and brain function in birds and mammals is reviewed by Rattenborg et al. (2009) and generally discussed in Linden (2007).

The laterality of the bird brain—the tendency of either the right or left hemisphere to sense and integrate certain types of information—is discussed by Ortega et al (2008) and Izawa et al. (2005).

The cognitive basis of tool use is discussed generally by Emery and Clayton (2009) and specifically for New Caledonian crows by Hunt et al. (2001) and Wier et al. (2004).

References Made within Chapter 3: Language

Kevin Smith's original recounting of the Missoula dog story was published as a letter in the *Montanan,* fall 2005. Subsequent discussions with Kevin filled in the details that we report.

The report of the talking crow from France was provided by bonjourfrederique@hotmail.com in response to an interview with Marzluff on the *Diane Rehm* show, October 2005.

Lorenz's tale of the speaking crow as well as Roah his raven is detailed in *King Solomon's Ring* (Lorenz 1952).

Esther Woolfson (2008) reports on her life with Spike and other corvids.

Macrobius's recounting of Caesar's encounters with talking ravens and the Romans who trained the birds is from *Saturnalia,* Book 2, 29, and 30 (Kaster 2011). We were alerted to the emperor Caesar's encounters with talking ravens by Fred Lohrer, Archbold Biological Station, and led to Kaster's book by Professor Catherine Connors, University of Washington.

Gretchen Copland told us about the raven at the Denver Zoo (Joe) who recognized her former owner (Bob).

Observations of speaking ravens at the Tower of London are from Sax (2011).

Pliny the Elder discusses talking birds in his *Natural History,* Vol. 10. (We read the translation by Bostock and Riley from 1865, which is online at www.perseus.tufts.edu/). While others suggest Pliny noted that a bird's tongue must be split to enable speech, we could not find this incorrect and cruel suggestion in his writing.

The syrinx of a bird is generally described by Goller and Larsen (1997) and Gill (2007) and that of a crow specifically is described and illustrated by Chamberlain et al. (1968).

The ear of a bird and its hair cells is described in Whittow (2000), Köppl et al. (2000), and Theunissen and Shaevitz (2006). The hearing range of birds (up to 7 kHz in crows, 15 kHz in starlings) and people (from 20 Hz to 20 kHz) is from Edwards (1943) and Welty and Baptista (1988).

Wernicke's area in the human brain is discussed by Allen (2009). The anatomy and learning of birdsong and human language is compared by Jarvis (2004). The ability of birds and humans to remember auditory information is from Kretzschmar et al. (2008).

Vocal learning by bats and early bat fossils is from Jepsen (1966) and Knörnschild et al. (2010). Recent findings suggesting that elephants are also capable of vocal learning are from Poole et al. (2005).

Parrot evolution is based on the genetic work of Wright et al. (2008).

General summary of song learning is from Whittow (2000), White (2001), Rowe and Skelhorn (2004), Beecher and Brenowitz (2005), and

Gill (2007). Human language and how animal communication differs is described by Hauser (2000), Anderson (2004), Holden (2004), Fitch and Hauser (2004), Kaplan (2004), Premack (2004), and Allen (2009).

Detailed descriptions of the brain centers and pathways for hearing, song learning, and singing are from Nottebohm (1984), Simpson and Vicario (1990), Shaw (2000), Köppl et al. (2000), Lavenex (2000), Okanoya et al. (2001), Jarvis (2004), Theunissen and Shaevitz (2006), Andalman and Fee (2009), and Gale and Perkel (2010). The term HVC is a proper name applied to a region in the nidopallium that is important to song learning. Reiner et al. (2004) trace its history as an original misnomer (Hyperstriatum ventrale pars caudale; the region is in the nidopallium, not hyperpallium) to the less objective "Higher Vocal Center," to simply the HVC.

The role of dopamine in bird social behavior including song is discussed by Goodson et al. (2009). Dopamine's role in song learning is particularly discussed by Gale and Perkel (2010).

The gene FoxP2 and its similar occurrence in human- and bird-language-learning centers is discussed by Scharff and Haesler (2005) and Allen (2009).

The function of REM sleep to song learning is demonstrated by Dave and Margoliash (2000).

Pepperberg's investigation of imitation and mimicry in her parrot Alex is from Pepperberg (2007).

Vocal mimicry in birds is reviewed by Garamszegi et al. (2007) and Kelley et al. (2008). Mimicry of hawk calls by corvids is reported by Hailman (1990) and Ratnayake et al. (2010).

Arbitrary, referential signaling in birds including corvids and parrots is from Hope (1980), Bugnyar et al. (2001), Hauser (2000), Kaplan (2004), Marzluff and Angell (2005a), Yorzinski and Vehrencamp (2009), and Giret et al. (2010).

Specific raven and crow calls are described and interpreted by Frings et al. (1958), Thompson (1968), Chamberlain and Cornwell (1971), Richards and Thompson (1978), Thompson (1982), Conner (1985), Kilham (1985, 1986), Heinrich and Marzluff (1991), Gorenzel and Salmon (1993), Hauser and Caffrey (1994), Marzluff and Angell (2005a), Yorzinski et al. (2006), Yorzinski and Vehrencamp (2009), Marzluff and Marzluff (2011).

Mimicry by mated birds of each other's calls is discussed by Thorpe and North (1966). A review of mimicry by songbirds is provided by Garamszegi et al. (2007). Use of learned songs in group identification by crows was studied by Brown (1985) and Brown and Farabaugh (1997).

References Made within Chapter 4: Delinquency

Windshield-wiper thievery by ravens was reported by Cathi (Jones) Winings, Katharine Stieg, Regina Rochefort, David Williams, Mark Young, Bill McCuaig, Cori Conner, and Mason Reid. A short news piece on the events in the North Cascades and elsewhere was reported by Wootton (2006). Kilham (1989) reports windshield-wiper destruction by a pet raven. Johan Peleman wrote to us in September 2011 to report on crows in southern Netherlands who stole windshield wipers during the nesting season.

How the amygdala, hormones, and dopamine work to condition learned fears is discussed by Stewart et al. (1996), Blair et al. (2001), McGaugh (2004), Maren (2005), Phelps (2006), and Diaconescu et al. (2009).

Destruction of structures at China Lake Naval Air Weapons Station was investigated by Boarman et al. (2002).

You can learn more about the Kea by viewing the film at www.nzonscreen.com/title/kea-mountain-parrot-1993.

The general neurobiology of associative learning during foraging is explored by Sherry and Mitchell (2007) and Adams-Hunt and Jacobs (2007).

The role of dopamine in associative learning is complex and often argued. In 2005 a large review article and detailed responses from a wide range of perspectives was published in the journal *Behavioral and Brain Sciences* (Depue and Morrone-Strupinsky 2005). In 2007 a series of articles in *Psychopharmacology* also explored the changing concepts concerning the role of dopamine (Salamone 2007, Berridge 2007). Other recent reviews include Schultz (2006), Grace et al. (2007), and Goodson and Kabelik (2009).

Reports of mischievous corvids are as follows: Australian ravens and forest ravens killing lambs (Rowley 1969), carrion crows stealing soap and candles (Higuchi and Kurosawa 2010), American crows eating piping plovers (Bragg 2010), common ravens eating desert tortoises (Boarman 2003), common ravens preying upon sage grouse (Bui et al. 2010), corvids rolling back sod to get grubs in Menlo Park, California (reported to us by Amy Hall on September 30, 2008), liver popping by crows (Surman 2005), common ravens preying on eared grebes (Yellowstone Science 2004).

American crows fishing for bass, stickleback, and salmon smolts in Lake Washington was reported by Professor Jim Kenagy, University of Washington in June 2009, and a photo of such appeared in the *Seattle Times,* June 23, 2006. Evan C. Evans III reported on the Clear Lake, California, crows fishing. Dorothea Forrest and Elwin Wright reported swimming ravens in Grants Cove, Alaska.

Ravens crushing vole tunnels was described by Mallory (1968).

Noni Pope told us about ravens and robins in 2006, Donna Darm reported the same for crows in Seattle in 2007, and Katie Roloson documented the hunting of crossbills by Elvis in the North Cascades of Washington. We have studied Elvis for four years, and he is starting to open the zippers on packs and dive into open car windows to grab a quick meal. As we try to stifle these activities, we hope he doesn't start to steal windshield wipers. David and Cindi Sonntag told us about a jungle crow stealing their dog's food in Motoyoyogi, Japan, in 2006. We learned about the luring crow in eastern Washington from Jack Ingram, and the luring raven in Alaska from Sandy Harbanuk.

Corvid-pair bonds are discussed by Marzluff and Angell (2005a). Their cooperative hunting was reported by Jeff Williams in 2007 and by Bill Harrower from the University of British Columbia in 2009. Ravens hunting seals and ganging up to steal from sled dogs is reported in Bent (1946). Ravens hunting porcupines is from Gehring (1993). Crows stealing from otters was observed by Kilham (1982). The experimental work with rook cooperation was done by Seed et al. (2008).

Smoking house crows were observed by Judie Ellis, and a photograph was published in Telegraph.co.uk on 19 May, 2010. Influence of nicotine on chickens is from Marley and Seller (1974). Jay Bennett and his family told us of their former pet crow in August 2010. Bob Armstrong shared a photo of a raven drinking a cup of Raven's Brew coffee in a Juneau parking lot with us in September 2008. Noni Pope reported the beer-drinking ravens in 2006.

Beverly McCann told us in 2005 about her clothespin-stealing crow in Michigan. Darla Dehlin told us about her raven troubles on March 16, 2009. Daryl Clark recounted the apple attack by a crow in 2007. A visitor to the Leavenworth, Washington, bookstore recounted his experiences with crows while golfing in 2005. Lela Brown told us in 2006 about the crows and kayakers in Barclay Sound.

Delbert Wichelman (2006) reported on his childhood pet crow returning his sister's ring.

Suzanne Wyman told us of her experiences with crows in British Columbia during an interview in June 2010.

Headlines from newspapers appeared in the *Los Angeles Times*, Associated Press (Twin Falls, Idaho), and Spiegel online.

Statistics on United States' killing of crows and ravens is from the 2009 annual report of the USDA Wildlife Services, www.aphis.usda.gov/wildlife_damage/prog_data/prog_data_report_fy_2009_disclaimer.shtml.

Management strategies for corvids are from Bui et al. (2010) and Boarman (2003). Dr. Richard Golightly from Humboldt State University is researching the effectiveness of sickening eggs to deter nest predation by corvids on murrelet eggs. Application of this process on crows, generally called conditioned taste aversion, is provided by Nicolaus and Cassel (1983) and discussed generally also at www.conditionedtasteaversion.net/.

Cape Cod headline is from April 24, 2010, Capecodonline.com (Bragg 2010).

References Made within Chapter 5: Insight

Heinrich's string-pulling experiments are discussed in Heinrich (1995, 2000). The tests of string pulling, both aided and unaided by mirrors, are from Taylor et al. (2010a).

Rooks' use of stones to raise water was studied by Bird and Emery (2009a). In a critique of this article, Russell Gray and others suggested the possibility of immediate reinforcement rather than insight as the mechanism emulating the rooks' behavior. Their use of hooks was reported by Bird and Emery (2009b).

The abilities of animals to quickly learn and strongly retain ecologically important associations is reviewed by Domjan et al. (2004), and Domjan (2005).

The three-stage puzzle solved by New Caledonian crows was reported by Taylor et al. (2010b).

Time-traveling scrub-jays that plan for breakfast was reported by Raby et al. (2007). Clayton and Dickinson (1998) and Clayton et al. (2001a, b) also investigate general planning by scrub-jays. Atance and O'Neill (2001) and Crystal (2010) review how people and animals use their brains to plan for the future.

Raven natural history is described by Heinrich (1999) and Marzluff and Marzluff (2011). Kilham's observations of possible deceptive caching are described in Kilham (1989). Jimmy the crow's understanding of dog knowledge is from Straub (2007). Valerie Allmendinger told us (and a radio audience on CBC's radio program *Wise Guys* in 2010) about her experience with crows and geese in Canada in 2006. Experiments on deceptive caching by ravens were conducted by Heinrich and Pepper (1998), Bugnyar and Kotrschal (2002), and Bugnyar and Heinrich (2005). Scrub-jay deception is discussed by Clayton et al. (2001a). More general discussion of the theory of mind can be found in Heyes (1998).

Jene Jernigan wrote us in 2007.

Bugnyar et al. (2004) and Schloegl et al. (2007) studied how ravens used the gaze of people to locate hidden foods.

Many people have told us about how crows count hunters and discriminate among people with guns and brooms, but W. Cone Johnson, MD, dug back into his class records to report the details of the North Texas experiments.

Gary Clark wrote us about his unusual experience with a crow March 27, 2006. We visited him and his wife on November 7, 2008. Reports of gifting from Nancy and Leona were recorded on the call-in portion of the *Diane Rehm* radio show October 25, 2005. Gayle Bodorff recounted her gift of a red-and-white toy bomb on April 22, 2006. Molly LeMaster spoke with us in 2007, Barbara Arnold told us her story in 2008, and Beth experienced a key gift in 2006. Janet Brandt contacted us in 2009. Eric Freeman contacted us in 2007.

Judith and Russell Balda confirmed the details of their experience with the hanging crow in October 2011. Mike Przystal reported on November 28, 2011, a similar experience with a young magpie that was stuck in his wooden fence in Calgary, Alberta. Mike received no gifts from the rescued magpie, but when he later encountered the magpie and its family they called softly and were not alarmed. A few weeks later, while sitting inside the house a young magpie, likely the one Mike rescued, spotted him and tried to get in the window.

Insights into gifting by dolphins was relayed to us by Janet Mann and is described in Dudzinski and Frohoff (2008). Condor and bowerbird attraction to shiny objects is from Endler and Day (2006) and Mee et al. (2007). Evon Zerbetz told us of the use of human construction materials in nests of ravens from Alaska's North Slope in 2007.

Ben Jonson's comedy *Volpone; Or the Fox* was first produced in 1606 and brought to our attention by Arne Zaslove. Marzluff and Angell (2005a, b) discuss Northwest raven myths.

Lois Magnusson spoke with us about the use of a plate by crows in 2006.

References Made within Chapter 6: Frolic

Emilie and George Rankin recorded their observations of ravens in Rocky Mountain National Park during a trip in 1999 and relayed the information to us in 2010.

The ethological consideration of play and its definition is from Burghardt (2005). The chapters in that book as well as those in Bekoff and Byers (1998) consider a range of topics and examples of animal play from a scientific perspective. These books provided the basis for our conclusions that many animals play. More specific consideration of play by birds is from Thorpe (1966) and Ficken (1977).

Play in jungle crows (diving into wind, hanging on wire, and stuffing dung in deer ears) is from chapters in Higuchi and Kurosawa (2010).

Sliding by ravens and crows is from Gilbert (1988), Moffett (1984), and Heinrich and Smolker (1998). In 2008, in Juneau, Alaska, Bob Armstrong described his observations of ravens sliding on their backs. During that conversation we heard repeated descriptions of ravens sliding on roots and banks in various Alaska locations. Some sliding ravens caused snowballs to form, which they picked up and carried. You can watch a short video of ravens sliding down a snowy roof in Yellowstone on YouTube, www.youtube.com/watch?v=LpsDWFei9Z4 (accessed June 22, 2011). A video of a hooded crow sledding on the lid of a plastic container can be seen on YouTube, www.youtube.com/watch?v=mRnI4dhZZxQ (accessed February 7, 2012).

Mary Palm reported springing crow play to us in 2006. Donna Winter's crow, who adopted her in 1970, is chronicled in *Year of the Crow* (Winter 2005). Richard Borgen reported on wire-riding ravens in 2010.

Raven tug-o-war is described in Heinrich and Smolker (1998) and Marzluff and Marzluff (2011). Chasing play by choughs is reported in Ficken (1977). Thorpe (1966) reports on the object play by a raven at the Copenhagen Zoological Gardens. Woolfson (2008) details her and her dog's play with Spike the magpie.

Carole Anne Coffey reported her playful Omar the crow to us in 2009. An interview with her daughter, Juliana, in February 2011 elaborated many aspects of play by their pet crow.

Kilham (1989) reports on crows chasing turkeys.

The connections between life history strategies and play are discussed by Ricklefs (2004).

The learning and behavioral benefits of play are reviewed by Panksepp et al. (1984), Siviy (1998), and Trezza et al. (2010).

The neurobiology of play is reviewed by Siviy (1998). The role of endorphins to reward and dopamine to motivate a variety of social and rewarding behaviors, including play, is discussed by Vanderschuren et al. (1995), Piazza and Le Moal (1997), Moles et al. (2004), Depue and Morrone-Strupinsky (2005), Barbano and Cador (2007), Salamone et al. (2007), Berridge and Kringelbach (2008), Vanderschuren et al. (2008), and Berridge et al. (2009).

The social brain network idea is from Insel and Fernald (2004), Goodson (2005), and Goodson and Kabelik (2009). The locations of dopamine and cannabinoid receptors in songbird brains is mapped by Kubikova et al. (2009) and Soderstrom and Johnson (2000). The important steroid produced by the preoptic area is vacotocin.

The important role of dopamine and its influence on synapses in the nucleus accumbens to affect social behavior and goal-directed, associative learning, mostly in mammals including humans, is reviewed by Depue and Morrone-Strupinsky (2005), Aragona et al. (2006), Iordanova et al. (2006), Björklund and Dunnett (2007), Grace et al. (2007), Salamone et al. (2007), and Schultz (2007). The particular role dopamine, opioids, and the social brain network in play is reviewed by Siviy (1998), Burghardt (2005), and Trezza et al. (2010). The connection between dopaminergic neurons in the midbrain (central gray, ventral tegmental area and substantia nigra) and striatum (Area X in particular for songbirds, but also the nucleus accumbens generally) in birds as a homologous system to that found in mammals is discussed by Stewart et al. (1996), Butler and Cotterill (2006), Gale and Perkel (2006, 2010), and Goodson et al. (2009).

How opioids respond to stress and fear and then affect play is discussed by Siviy et al. (2006). The general control of social behaviors, including play, by opioids is from Panksepp et al. (1984), Vanderschuren et al. (1995, 2008), and Trezza et al. (2010).

The cannabinoid system and its effects on learning and social behavior, including play, is reviewed by Elphick and Egertová (2001), Trezza and Vanderschuren (2008), and Lutz (2009). Cannabinoid receptors in birds are reviewed by Soderstrom and Johnson (2000).

The excitatory neurotransmitters include glutamate, and a common inhibitory neurotransmitter is gamma aminobutyric acid (GABA).

The complex interaction among opioids, dopamine, endocanabinoids, and hormones that affect social behavior, including play, is reviewed by Siviy (1998), Burghardt (2005), Depue and Morrone-Strupinsky (2005), and Trezza et al. (2010). Some effects of serotonin and dopamine in birds are reviewed by van Hierden et al. (2002).

The concept of neural ensembles is from Depue and Morrone-Strupinsky (2005).

Panksepp et al. (1984) elaborate the many functions of play and especially how important play is to the development and refinement of social skills.

References Made within Chapter 7: Passion, Wrath, and Grief

Many people have sent us their observations of crow and raven gatherings around dead conspecifics. Deborah Raymond made her observation on November 1, 1999. Rod Stephens's observation was reported in the Fairbanks, Alaska, *Daily News-Miner* on November 22, 2009, and Craig Fritch told us a story in 2006 but made the observation between 1989 and 1995. Kay Schaffer took photographs of the crow surrounded by sticks and detailed the funeral at her home in August 2010. She shared her observations with us in September 2011 and followed up by questioning those who lived around her about the sticks. No one claimed responsibility for placing the sticks, which were obviously arranged. Bekoff (2009) reports on another instance when corvids, in this case magpies, gathered around a dead flock mate and brought bits of grass to the site.

The role of the amygdala in evaluating social and emotional significance of sensory information is discussed by LeDoux (2000), Blair et al. (2001), McGaugh (2004), Goodson (2005), Samson and Paré (2005), Phelps (2006), LeDoux and Phelps (2008), Wager et al. (2008), Allen (2009), and Bliss-Moreau et al. (2010). The response of a quail's amygdala to fearful stimuli was determined by Saint-Dizier et al. (2009). We suspect the amygdala links with the sensory Wulst, the entopallium in birds that are looking at a dangerous setting to increase perception. These regions are closely linked through the social brain network, notably wired in with the hippocampus, nucleus accumbens, striatum, and thalamus. The specific interaction of stress hormones and opioids on synapses in the amygdala and with other brain regions is suggested by McGaugh (2004). Learning about fear is accomplished by building strong, easily fired synapses in the nucleus accumbens between neurons from the amygdala, the hippocampus, and other forebrain regions.

The cellular basis of memory is reviewed by Rose and Stewart (1999), Blair et al. (2001), Insel and Fernald (2004), Linden (2007), and Kandel (2009). The role of dopamine is summarized by Depue and Morrone-Strupinsky (2005), Grace et al. (2007), Schultz (2007), and Salamone et al. (2007).

The influence of grief on the human brain is discussed by Freed et al. (2009) and O'Connor et al. (2009). The brain structures affected by sadness are studied by Goldin et al. (2005) and Wang et al. (2005).

Grief in primates is discussed by Barbara King (2010, and in a forthcoming book tentatively titled *How Animals Grieve*). Various relations among pair-bonded birds are described by Heinrich (2011).

The biology of pinyon jays is from Marzluff and Balda (1992).

The neurobiology of pair bonding in voles is summarized by Young and Wang (2004) and Aragona et al. (2006). When voles associate the opiate reward of sex with a specific female, it can be thought of as forging a neural ensemble under the influence of dopamine, hormones, and opioids in the nucleus accumbens that links neurons from a discriminating amygdala and with others throughout the social brain.

The vasotocin system of birds is discussed by Jurkevich et al. (2001).

David Caley observed two crows helping to support an injured crow in August 2008 on Two Hundred Thirty-Fifth Place, Sammamish, Washington.

As with crow funerals, we have received several observations of crow murders. We discussed some in Marzluff and Angell (2005a), but here report fresh observations by Carol Strickland in 2006, Julie Lawell in 2007, Kenneth Geiersbach in 2009, and Barbara Brozyna in 2006.

Goodson (2005) and Goodson et al. (2009) discuss how the social brain responds to aggression. The role of opioids in reducing aggression in birds is from Kotegawa et al. (1997).

Madeleine Kornfield reported her observation of ravens using weapons against a great horned owl in 2010. Russ Balda (2007) published his observation of jays and crows fighting with a stick. The use of stones as weapons by corvids is discussed by Montevecchi (1978).

Scientists refer to interactions after conflicts that involve other animals as "third-party affiliations." Technically, consolation is a type of third-party affiliation that is not initiated by the third party and one in which stress is reduced by the affiliation. In birds it is known that third parties (as well as aggressors and victims) initiate interactions in ravens as discussed by Fraser and Bugnyar (2010, 2011) and rooks as studied by Seed et al. (2007). Recent work suggests that such affiliation lowers stress in rooks (Corina Logan, personal communication August 13, 2011). Logan also expanded the work of Seed et al. (2007, 2008) to include Eurasian jays (manuscript submitted to *Animal Behaviour*;

personal communication). Woolfson (2008) describes being consoled by her pet magpie.

The examples, from many we receive, of metaphysical interactions with crows include reports from India by Padmini Pooleri in 2010, Portland by Josh Mong in 2007, Tacoma by Mary Coggins in 2010, Seattle by Jim Shumacher in 2006, New Marshfield, Ohio, by Crow Swimsaway in 2006 (see also www.church-of-earth-healing.org/ for more information), Seattle by Kent Bush in 2005, California by Cathleen Handlin in 2011, Ketchikan by Deloris Churchill in 2005, Vancouver Island by Suzanne Wyman in 2010, and Port Townsend by Dana Casey in 2009. Published accounts of metaphysical interactions with ravens include Abram (2010) and Sax (2011).

Benjamin Libet's research is synthesized in his 2004 book.

References Made within Chapter 8: Risk Taking

The rise in bald eagle populations is chronicled in Watts et al. (2008).

The biology of predator mobbing is discussed in Goodwin (1986), Kilham (1989), Ratcliffe (1997), Marzluff and Angell (2005a), and dos Anjos et al. (2009). Similarities between barking and mobbing are drawn by Lord et al. (2009).

The ability of birds to learn about dangerous places by observing the actions of others is suggested by Griffin and Boyce (2009).

Glenn Rimbey was reading while his son drove over the raven in 2006. He was jarred to look back as his son yelled, surely thinking the raven was dead, but it simply resumed eating until the next car again forced it to squat low on the road. Glenn told us about this incident in 2008. We learned of another similar incident in 2011 as Gail Olsen was driving about twenty miles an hour on the afternoon of October 16 along a well-traveled Seattle road. As she approached three crows eating in the roadway, two flew off, but the third remained in place and ducked as the car passed over the top of the bird and its meal.

Robert Mooney reported his observations of crow habituation to traffic along the roadways of Ohio to us in July 2009. The behavior of Japanese carrion crows walking and cracking nuts in crosswalks can be viewed on the PBS program "The Life of Birds," www.pbs.org/lifeofbirds/brain/index.html. The science behind the nutcracking antics of Japanese crows is detailed in Nihei and Higuchi (2001).

Matt Betts told us about the apparently suicidal crow in 2008. Rita Ross wrote about another possible suicide in October 2011. In that case, an injured crow was aside a road with a broken wing. As Rita approached, the bird hopped away and down the road toward an intersection where it appeared to jump in front of a small truck's front tire and was killed. In this case, it is hard to tell if the bird was aware of the truck, but it raises the possibility that injured birds become reckless and susceptible to other sources of mortality. Wright and Spence (1976) demonstrated the resultant lack of fear in pigeons that had damaged amygdalas. Saint-Dizier et al. (2009) refined this claim and showed that damage to the back portion (anterior) of the amygdala reduced fear response in Japanese quail. Llinás and Steriade (2006) link the firing patterns of neurons that connect the thalamus with the forebrain to vigilance.

Dingemanse et al. (2004) related the personalities of great tits to their survivorship. On October 25, 2005, a caller to the *Diane Rehm* show told us about the crow reactions to high-speed trains in New Hampshire.

The detrimental effects of cats on birds are reviewed by Hildreth et al. (2010). The influence of cats on endangered crows can be found in the U.S. Fish and Wildlife Service recovery plans for the Hawaiian crow and the Mariana crow (available at www.fws.gov/pacificislands/recoveryplans.html).

Porter Evans (2006) writes about his cat's interactions with crows.

The Collito video is available on YouTube.com by searching for "kitten and crow."

Jeanette Griver (2004) writes about her sheepdog's antics with a crow.

The photo of the dog offering a play bow to a crow is from Hoeschele (1989).

Mech (1970) reports on ravens flying low over wolf packs. Today, observers in Yellowstone National Park often witness the interactions between ravens and the reintroduced wolves that cruise Yellowstone's Northern Range.

Mary Ronback told us about being adopted by a raven, and she sent photos of the bird perched on Jim's arm in 2010. Lydia Janik relayed her interactions with tame crows to us in 2010. Jack Fellman's experience was written as a letter to the magazine *Birds and Blooms* (October-November, 2006).

The neurobiology explaining the motivation of social animals to seek their own kind is from Panksepp et al. (1978). Depue and Morrone-Strupinsky (2005) and Grace et al. (2007) review the neurobiology of goal-directed behaviors. The interaction of hormones with opioids and dopamine in affiliative, and risky behavior is discussed by van Hierden et al. (2002) and Piazza and Le Moal (1997). Young and Wang (2004) discuss the reward circuits of bonding. The importance of the amygdala to proper social behavior is discussed by Bliss-Moreau et al. (2010). The ventral tegmental area is an important part of the reward and motivation circuit we discuss. It is from this midbrain region that neurons that release dopamine arise. Together with the nucleus accumbens (and striatum in general) the ventral tegmental area, septum, and pallidum form key connections in the reward and motivation circuit relevant to social behavior like bonding.

Salzen et al. (1979), Horn and colleagues (1981, 1983, 1984), and Insel and Fernald (2004) discuss the neurobiology of imprinting. In birds it is the medial mesopallium that is important to synapses formed during imprinting.

References Made within Chapter 9: Awareness

Susanne Drakborg first told us of her parents' interactions with magpies in Västerås, Sweden, in August of 2008. Subsequent interviews in July and August 2010, including discussions and interviews with Inda and Abbe, confirmed details.

Larry Makinson, Carl Marti, Prisca Cushman, Jingle Ruppert, and Gene Carter each contacted us with their stories in August 2008. Shelly and Jim Leonard wrote us in October 2009. James Anderson wrote us in January 2010. We met Phyllis Alverez in 2003; her story also appeared in the *Seattle Times* on November 24, 2003.

Lawrence Kilham (1989) describes his interaction with a raven in Iceland.

Marzluff et al. (2010) report on the details of the facial recognition experiments with Seattle crows. Spear (1988) also used masks to conceal his identity from the gulls he was studying.

The information content of corvid alarm calls is discussed by Brown (1985), Marzluff and Balda (1992), Marzluff and Angell (2005a), Yorzinski et al. (2006), and Yorzinski and Vehrencamp (2009). Siberian jay scolding is described by Griesser (2008).

Midbrain centers for vocalizations like alarm calls are described by Seller (1981) and Shaw (2000). The function of the optic tectum is from Whittow (2000) and Watanabe and Masuda (2010). Riters and Ball (1999) also suggest that the midbrain (specifically the medial preoptic nucleus of the hypothalamus) is important to controlling courtship behavior including song. The nucleus intercollicularius receives input from the forebrain song-learning circuits via the arcopallium as well as from the medial preoptic area of the hypothalamus, and perhaps substantially from the optic tectum. The tectum is layered, and its outer neurons, those first stimulated by a sight, convey visual information only, but the inner layers integrate sight, sound, and touch sensations. In pigeons diverse sensory information is either integrated in the optic tectum or in the nidopallium of the forebrain (Watanabe and Masuda 2010).

The hormone testosterone stimulates many aggressive and sexual actions in birds and mammals. In songbirds, testosterone is converted into estrogen in (among other areas) the preoptic nucleus of the hypothalamus, where it influences courtship behaviors such as nest building and singing (Riters et al. 1998, and Riters and Ball, 1999). The role of testosterone in mobbing by a corvid is discussed by Cully and Ligon (1986).

Potential involvement of the forebrain and midbrain in avian alarm calling is discussed by Kaplan (2008).

Neurobiology of discrimination is discussed by Phillmore (2008). The distinct role of the entopallium and the Wulst is reported by Watanabe et al. (2011).

The recognition of people by animals is reviewed by Kendrick and Baldwin (1987), Tate et al. (2006), Dahl et al. (2007), and Marzluff et al. (2010). It should be noted that species other than crows can have been shown in the lab to learn to recognize people; crows, however, spontaneously do so in the wild. How people recognize the faces of others is discussed by Ashwin et al. (2007), Eimer and Holmes (2007), Gobbini and Haxby (2007), Palermo and Rhodes (2007), Rolls (2007), and Vuilleumier and Pourtois (2007). Generally we use a core recognition system dispersed among at least three regions in our sensory cortex (the superior sulcus, the inferior occipital gyrus, and the fusiform gyrus) and networked with two extended systems that convey the history and emotional significance of the person. Knowledge about the person includes information about an individual's traits, attitudes, and mental states, which are deciphered by our superior temporal sulcus, biographical knowledge held in our anterior temporal cortex, and episodic memories from our precuneus and posterior cingulate. Emotion is conveyed by our amygdala, insula, and the reward system of the striatum.

Our research on the reaction of crows to our gaze (directed or averted) and expression (smiling versus frowning) was led by Dr. Barbara Clucas (manuscript in preparation). Recognition of human faces by pigeons, even inverted ones, is discussed by Jitsumori and Makino (2004), Cook et al. (2005), and Watanabe and Masuda (2010). The inability of humans to recognize inverted faces is from Yin (1969) and Parr et al. (1998). The possibility that recognition by chicks is controlled by the brain's right hemisphere is from Vallortigara and Andrew (1994).

Van der Linden et al. (2009) and Axmacher at al. (2009) review the use of various types of imaging to investigate the activity of animal brains. PET scanning is described by Allen (2009).

The brain-stem relay that was activated in our PET study by crows seeing a dangerous face was the nucleus reticularis pontis caudalis, known to be stimulated by the amygdala in response to the sight of a fearful object in mammals. The role of the amygdala to fear responses in Japanese quail is from Saint-Dizier et al. (2009). This study suggested that the arcopallium or posterior pallial amygdala was important to a quail's fear response, while our study suggested that in crows it is the subpallial amygdala, the septum, and the lateral part of the bed nucleus of the stria terminalis that is activated by fear.

Rapid learning and persistent memory of danger is reviewed by Griffin (2004), Sugai et al. (2007), and Marzluff et al. (2010).

Social learning about danger is described by Curio et al. (1978), Kavaliers et al. (2003), Griffin (2004), Olsson and Phelps (2007), Olsson et al. (2007), and Bruchey et al. (2010). More general discussion of social learning can be found in Heyes (1994), Lefebvre (1995), Langen (1996a, b), Whiten (2000), and Champagne and Curley (2005). Our research on social learning of a dangerous person by crows is detailed in Cornell et al. (2012).

Mirror neurons are discussed in Rizzolatti and Craighero (2004), Iacoboni (2005), Rizzolatti and Fabbri-Destro (2008), and Prather et al. (2008).

Stories about continued harassment of people by crows are common. Randy Gerber wrote us in 2008 and Katrina Anderson wrote us in 2009 to share their stories. Stories about forgiveness are less frequent. Donna Barr rescued her crow in the 1980s and wrote us in 2010 to relay her experience. Bill Romaniuk told us his story from 1991 in 2010. Ted Hayes relayed his story to us in 2010, while Nancy Kool discussed her experience with us in 2009.

General neural changes during fear learning are reviewed by Kim et al. (2006), Delgado et al. (2008), and Reinhardt et al. (2010). The more specific role of the amygdala in learning about fearful experiences is reviewed by Weisskopf et al. (1999), LeDoux (2000), Blair et al. (2001), McGaugh (2004), Maren (2005), Phelps (2006), LeDoux and Phelps (2008), and Öhman (2008). The interactions of hormones, dopamine, and opioids in learning during fearful experiences is discussed by Martin et al. (1979), Stewart et al. (1996), Korte (2001), Iordanova et al. (2006), and Diaconescu et al. (2009). Forgetting fearful experiences is reviewed by Lee and Kim (1998), Santini et al. (2001), Milad and Quirk (2002), and Lissek and Güntürkün (2003).

Self-recognition in magpies was demonstrated by Prior et al. (2008). The Great Falls, Montana, *Tribune* published a photo of a raven reacting to its reflection on March 1, 2007. Both Grant Stevenson from Fountain Hill, Pennsylvania, and Jane O'Neil from Ravensdale, Washington, report similar responses from crows to their reflections.

References Made within Chapter 10: Reconsidering the Crow

Striedter (2005) reviews the evolution of brains in vertebrates. Butler and Cotterill (2006) compare the neural bases of consciousness between birds and mammals.

The connections between innovation, general lifestyles, and large brains have been made by Wyles et al. (1983), Lefebvre et al. (1997, 2004), Iwaniuk and Nelson (2003), Marzluff and Angell (2005a), Sol et al. (2005, 2007), and Lefebvre and Sol (2008).

The exceptionally large brain of the New Caledonian crow is reported by Cnotka et al. (2008) and generally discussed by Emery and Clayton (2004b) and Rogers (2004).

Ricklefs (2004) and Sol et al. (2007) explore the relationship between sociality, lifespan, and brain size.

Wyles et al. (1983) and Lefebvre et al. (2004) demonstrate the link between morphological evolution and brain size.

The role of the NCL and CDL as the bird brain's executive center is discussed by Güntürkün (2005), Rose and Colombo (2005), Butler and Cotterill (2006), and Kirsch et al. (2009).

Heyes (1998) discusses the ability of nonhuman animals to possess a theory of mind.

Marzluff and Angell (2005a, b) discuss the cultural coevolution that we feel exists between humans and corvids.

Ken Botwright called us in 2011 to tell of his family pet crow and the ways in which it affected his childhood on the plains of Ontario.

Cal Aylmer wrote about her deep connection to crows beginning in April 2009. Denise Storm told us of her cheer in watching ravens in August 2010. Levi Fuller produced an album, "This Murder Is a Peaceful Gathering," in 2007. Jeanne Shepard writes the cartoon column "Urban Crow" from Seattle.

References

Aboitiz, F. 2011. Genetic and developmental homology in amniote brains: Toward conciliating radical views of brain evolution. *Brain Research Bulletin* 84:125–36.

Abram, D. 2010. *Becoming Animal: An Earthly Cosmology*. New York, NY: Pantheon Books.

Adams-Hunt, M. M., and L. F. Jacobs. 2007. Cognition for foraging. Pages 105–38. In *Foraging: Behavior and Ecology,* edited by D. W. Stephens, J. S. Brown, and R. C. Ydenberg. Chicago, IL: University of Chicago Press.

Allen, J. S. 2009. *The Lives of the Brain*. Cambridge, MA: Harvard University Press.

Alvarez-Buylla, A., and F. Nottebohm. 1988. Migration of young neurons in adult avian brain. *Nature* 335:353–54.

Andalman, A. S., and M. S. Fee. 2009. A basal ganglia-forebrain circuit in the songbird biases motor output to avoid vocal errors. *Proceedings of the National Academy of Science, USA* 106:12518–23.

Anderson, S. R. 2004. *Doctor Dolittle's Delusion*: Animals and the Uniqueness of Human Language. New Haven, CT: Yale University Press.

Aragona, B. J., Y. Liu, Y. J. Yu, J. T. Curtis, J. M. Detwiler, T. R. Insel, and Z. Wang. 2006. Nucleus accumbens dopamine differentially mediates the formation and maintenance of monogamous pair bonds. *Nature Neuroscience* 9:133–39.

Ashwin, C., S. Baron-Cohen, S. Wheelwright, M. O'Riordan, and E. T. Bullmore. 2007. Differential activation of the amygdala and the "social brain" during fearful face-processing in Asperger Syndrome. *Neuropsychologia* 45:2–14.

Atance, C. M., and D. K. O'Neill. 2001. Episodic future thinking. *Trends in Cognitive Sciences* 5:533–39.

Avian Brain Nomenclature Consortium. 2005. Avian brains and a new understanding of vertebrate brain evolution. *Nature Reviews Neuroscience* 6:151–59.

References

Axmacher, N., C. E. Elger, and J. Fell. 2009. The specific contribution of neuroimaging versus neurophysiological data to understand cognition. *Behavioural Brain Research* 200:1–6.

Balda, R. P. 2007. Corvids in combat: With a weapon? *Wilson Journal of Ornithology* 119:100–102.

Balda, R. P., and A. C. Kamil. 1992. Long-term spatial memory in Clark's nutcrackers, *Nucifraga columbiana*. *Animal Behaviour* 44:761–69.

Balda, R. P., A. C. Kamil, and P. A. Bednekoff. 1996. Predicting cognitive capacity from natural history: Examples from four species of corvids. Pages 33–66. In *Current Ornithology,* edited by V. Nolan and E. D. Ketterson. New York, NY: Plenum Press.

Barbano, M. F., and M. Cador. 2007. Opioids for hedonic experience and dopamine to get ready for it. *Psychopharmacology* 191:497–506.

Basil, J. A., A. C. Kamil, R. P. Balda, and K. V. Fite. 1996. Differences in hippocampal volume among food storing corvids. *Brain Behavior and Evolution* 47:156–64.

Bates, L. A., and R. W. Byrne. 2007. Creative or created: Using anecdotes to investigate animal cognition. *Methods* 42:12–21.

Bauer, E. E., M. J. Coleman, T. F. Roberts, A. Roy, J. F. Prather, and R. Mooney. 2008. A synaptic basis for auditory-vocal integration in the songbird. *Journal of Neuroscience* 28:1509–22.

Bednekoff, P. A., and R. P. Balda. 1996. Social caching and observational spatial memory in pinyon jays. *Behaviour* 133:807–26.

Beecher, M. D., and E. A. Brenowitz. 2005. Functional aspects of song learning in songbirds. *Trends in Ecology and Evolution* 20:143–49.

Bekoff, M. 2009. Animal emotions, wild justice and why they matter: Grieving magpies, a pissy baboon, and empathic elephants. *Emotion, Space and Society* 2:82–85.

Bekoff, M., and J. A. Byers, eds. 1998. *Animal Play: Evolutionary, Comparative, and Ecological Perspectives*. Cambridge, UK: Cambridge University Press.

Benhamou, S., and B. Poucet. 1996. A comparative analysis of spatial memory processes. *Behavioural Processes* 35:113–26.

Bent, A. C. 1946. *Life Histories of North American Jays, Crows, and Titmice*. Washington, DC: *Smithsonian, US National Museum Bulletin* 191.

Berridge, K. C. 2007. The debate over dopamine's role in reward: The case for incentive salience. *Psychopharmacology* 191:391–431.

Berridge, K. C., and M. L. Kringelbach. 2008. Affective neuroscience of pleasure: Reward in humans and animals. *Psychopharmacology* 199:457–80.

Berridge, K. C., T. E. Robinson, and J. W. Aldridge. 2009. Dissecting components of reward: "liking," "wanting," and learning. *Current Opinion in Pharmacology* 9:65–73.

Bingman, V. P., and K. P. Able. 2002. Maps in birds: Representational mechanisms and neural bases. *Current Opinion in Neurobiology* 12:745–50.

Bingman, V. P., and A. Gagliardo. 2006. Of birds and men: Convergent evolution in hippocampal lateralization and spatial cognition. *Cortex* 42:99–100.

Bird, C. D., and N. J. Emery. 2009a. Rooks use stones to raise the water level to reach a floating worm. *Current Biology* 19:1410–14.

Bird, C. D., and N. J. Emery. 2009b. Insightful problem solving and creative tool modification by captive nontool-using rooks. *Proceedings of the National Academy of Sciences USA* 106:10370–75.

Björklund, A., and S. B. Dunnett. 2007. Dopamine neuron systems in the brain: An update. *Trends in Neurosciences* 30:194–202.

Blair, H. T., G. E. Schafe, E. P. Bauer, S. M. Rodrigues, and J. E. LeDoux. 2001. Synaptic plasticity in the lateral amygdala: A cellular hypothesis of fear conditioning. *Learning & Memory* 8:229–42.

Bliss-Moreau, E., J. E. Toscano, M. D. Bauman, W. A. Mason, and D. G. Amaral. 2010. Neonatal amygdala or hippocampus lesions influence responsiveness to objects. *Developmental Psychobiology* 52:487–503.

Boarman, W. I. 2003. Managing a subsidized predator population: Reducing common raven predation on desert tortoises. *Environmental Management* 32:205–17.

Boarman, W. I., S. J. Coe, and W. Webb. 2002. Development of aversion techniques to prevent equipment damage by common ravens (*Corvus corax*) at China Lake Naval Air Warfare Station. *Report to China Lake Naval Air Warfare Station*. San Diego, CA: U.S. Geological Survey.

Bond, A. B., A. C. Kamil, and R. P. Balda. 2003. Social complexity and transitive inference in corvids. *Animal Behaviour* 65:479–87.

Bostock, J., and H. J. Riley. 1855. *The Natural History*. Translation of Pliny the Elder's Tenth Book, available at Perseus Digital Library, edited by G. R. Crane. www.perseus.tufts.edu/.

Bragg, M. A. 2010. Cape crow poison plan killed. www.capecodonline.com/apps/pbcs.dll/article?AID=/20100424/NEWS/4240319/-1/news01, April 24, 2010.

Brown, E. D. 1985. The role of song and vocal imitation among common crows (*Corvus brachyrhynchos*). *Zeitschrift für Tierpsychologie* 68:115–36.

Brown, E. D., and S. M. Farabaugh. 1997. What birds with complex social relationships can tell us about vocal learning: Vocal sharing in avian groups. Pages 98–127. In *Social Influences on Vocal Development,* edited by C. T. Snowdon and M. Hausberger. Cambridge: Cambridge University Press.

Bruchey, A. K., C. E. Jones, and M. H. Monfils. 2010. Fear conditioning by-proxy: social transmission of fear during memory retrieval. *Behavioural Brain Research* 214:80–84.

Bugnyar, T., and K. Kotrschal. 2002. Observational learning and the raiding of food caches in ravens, *Corvus corax*: Is it "tactical" deception? *Animal Behaviour* 64:185–95.

Bugnyar, T., and B. Heinrich. 2005. Ravens, *Corvus corax,* differentiate between knowledgeable and ignorant competitors. *Proceedings of the Royal Society of London B* 272:1641–46.

Bugnyar, T., M. Stowe, and B. Heinrich. 2004. Ravens, *Corvus corax,* follow gaze direction of humans around obstacles. *Proceedings of the Royal Society of London B* 271:1331–36.

Bugnyar, T., M. Kijne, and K. Kotrschal. 2001. Food calling in ravens: Are yells referential signals? *Animal Behaviour* 61:949–58.

Bui, T-V. D., J. M. Marzluff, and B. Bedrosian. 2010. Common Raven activity in relation to land use in western Wyoming: Implications for Greater Sage-Grouse reproductive success. *Condor* 112:65–78.

Burghardt, G. M. 2005. *The Genesis of Animal Play: Testing the Limits*. Cambridge, MA: MIT Press.

Butler, A. B. 2008. Evolution of brains, cognition, and consciousness. *Brain Research Bulletin* 75:442–49.

Butler, A. B., P. R. Manger, B. I. Lindahl, and P. Arhem. 2005. Evolution of the neural basis of consciousness: A bird-mammal comparison. *BioEssays* 27:923–36.

Butler, A. B., and R. M. J. Cotterill. 2006. Mammalian and avian neuroanatomy and the question of consciousness in birds. *Biological Bulletin* 211:106–27.

Capuzzo, M. 1993. The raven of Poe's famous poem is a feather in Free Library's cap, *Philadelphia Inquirer*. June 13, 1993.

Chamberlain, D. W., and G. W Cornwell. 1971. Selected vocalizations of the common crow. *Auk* 88:613–34.

Chamberlain, D. W., W. B. Gross, G. W. Cornwell, and H. S. Mosby. 1968. Syringeal anatomy in the common crow. *Auk* 85:244–52.

Champagne, F. A., and J. P. Curley. 2005. How social experiences influence the brain. *Current Opinion in Neurobiology* 15:704–9.

Clayton, N. S., and A. Dickinson. 1998. Episodic-like memory during cache recovery by scrub jays. *Nature* 395:272–74.

Clayton, N. S., D. P. Griffiths, N. J. Emery, and A. Dickinson. 2001a. Elements of episodic-like memory in animals. *Philosophical Transactions of the Royal Society of London B* 356:1483–91.

Clayton, N. S., K. Yu, and A. Dickinson. 2001b. Scrub-jays (*Aphelocoma coerulescens*) can form integrated memory for multiple features of caching episodes. *Journal of Experimental Psychology Animal Behaviour Processes* 27:17–29.

Cnotka, J., O. Güntürkün, G. Rehkämper, R. D. Gray, and G. R. Hunt. 2008. Extraordinary large brains in tool-using New Caledonian crows (*Corvus moneduloides*). *Neuroscience Letters* 433:241–45.

Conner, R. N. 1985. Vocalizations of common ravens in Virginia. *Condor* 87:379–88.

Cook, R. G., K. Goto, and D. I. Brooks. 2005. Avian detection and identification of perceptual organization in random noise. *Behavioural Processes* 69:79–95.

Cornell, H. N., J. M. Marzluff, and S. Pecoraro. 2012. Social learning spreads knowledge about dangerous humans among American crows. *Proceedings of the Royal Society B*. 279:499–508.

Crystal, J. D. 2010. Episodic-like memory in animals. *Behavioural Brain Research* 215:235–43.

Cully, J. F., and J. D. Ligon. 1986. Seasonality of mobbing intensity in the pinyon jay. *Ethology* 71:333–39.

Curio, E., U. Ernst, and W. Vieth. 1978. Cultural transmission of enemy recognition: One function of mobbing. *Science* 202:899–901.

Dahl, C. D., N. K. Logothetis, and K. L. Hoffman. 2007. Individuation and holistic processing of faces in rhesus monkeys. *Proceedings of the Royal Society of London B* 274:2069–76.

Dave, A. S., and D. Margoliash. 2000. Song replay during sleep and computational rules for sensorimotor vocal learning. *Science* 290:812–16.

Delgado, M. R., J. Li, D. Schiller, and E. A. Phelps. 2008. The role of the striatum in aversive learning and aversive prediction errors. *Philosophical Transactions of the Royal Society of London B* 363:3787–3800.

Depue, R. A., and J. V. Morrone-Strupinsky. 2005. A neurobehavioral model of affiliative bonding: Implications for conceptualizing a human trait of affiliation. *Behavioral and Brain Sciences* 28:313–95.

Diaconescu, A. O., M. Menon, J. Jensen, S. Kapur, and A. R. McIntosh. 2009. Dopamine-induced changes in neural network patterns supporting aversive conditioning. *Brain Research* 1313:143–61.

Dickens, C. 1900. *Barnaby Rudge: A Tale of the Riots of "Eighty."* Boston, MA: DeWolfe, Fiske, and Company.

Dingemanse, N. J., C. Both, P. J. Drent, and J. M. Tinbergen. 2004. Fitness consequences of avian personalities in a fluctuating environment. *Proceedings of the Royal Society of London B* 271:847–52.

Domjan, M. 2005. Pavlovian conditioning: A functional perspective. *Annual Review of Psychology* 56:179–206.

Domjan, M., B. Cusato, and M. Krause. 2004. Learning with arbitrary versus ecological conditioned stimuli: Evidence from sexual conditioning. *Psychonomic Bulletin & Review* 11:232–46.

dos Anjos, L., S. Debus, S. Madge, and J. M. Marzluff. 2009. Family Corvidae. Pages 494–640. In *Handbook of Birds of the World, vol. 14,* edited by J. del Hoyo, A. Elliott, and D. Christie. Barcelona, Spain: Lynx Edicions.

Dudzinski, K. M., and T. Frohoff. 2008. *Dolphin Mysteries: Unlocking the Secrets of Communication*. New Haven, CT: Yale University Press.

Edwards, E. P. 1943. Hearing ranges of four species of birds. *The Auk* 60:239–241.

Eimer, M., and A. Holmes. 2007. Event-related brain potential correlates of emotional face processing. *Neuropsychologia* 45:15–31.

Elphick, M. R., and M. Egertová. 2001. The neurobiology and evolution of cannabinoid signaling. *Philosophical Transactions of the Royal Society of London B* 356:381–408.

Endler, J. A., and L. B. Day. 2006. Ornament colour selection, visual contrast and the shape of colour preference functions in great bowerbirds, *Chlamydera nuchalis*. *Animal Behaviour* 72:1405–16.

Emery, N. J. 2006. Cognitive ornithology: The evolution of avian intelligence. *Philosophical Transactions of the Royal Society of London B* 361:23–43.

Emery, N. J., and N. S. Clayton. 2004a. The mentality of crows: Convergent evolution of intelligence in corvids and apes. *Science* 306:1903–7.

Emery, N. J., and N. S. Clayton. 2004b. Comparing the complex cognition of birds and primates. Pages 3–55. In *Comparative Vertebrate Cognition: Are Primates Superior to Non-Primates?,* edited by L. J. Rogers and G. Kaplan. New York, NY: Kluwer Academic Press.

Emery, N. J., J. M. Dally, and N. S. Clayton. 2004. Western scrub-jays (*Aphelocoma californica*) use cognitive strategies to protect their caches from thieving conspecifics. *Animal Cognition* 7:37–43.

Emery, N. J., and N. S. Clayton. 2009. Tool use and physical cognition in birds and mammals. *Current Opinion in Neurobiology* 19:27–33.

Emery, N. J., A. M. Seed, A. M. P. von Bayern, and N. S. Clayton. 2007. Cognitive adaptations of social bonding in birds. *Philosophical Transactions of the Royal Society of London B* 362:489–505.

Enggist-Dueblin, P., and U. Pfister. 2002. Cultural transmission of vocalizations in ravens, *Corvus corax*. *Animal Behaviour* 64:831–41.

Evans, P. 2006. *The Well-Lettered Cat*. Santa Barbara, CA: Lines Rampant Press.

Farries, M. A. 2001. The oscine song system considered in the context of the avian brain: Lessons learned from comparative neurobiology. *Brain, Behavior and Evolution* 58:80–100.

Farries, M. A. 2004. The avian song system in comparative perspective. *Annals of the New York Academy of Science* 1016:61–76.

Feenders, G., M. Liedvogel, M. Rivas, M. Zapka, H. Horita, E. Hara, K. Wada, H. Mouritsen, and E. D. Jarvis. 2008. Molecular mapping of movement-associated areas in the avian brain: A motor theory for vocal learning origin. PLoS One 3:e1768.

Ficken, M. S. 1977. Avian play. *Auk* 94:573–82.

Fitch, W. T., and M. D. Hauser. 2004. Computational constraints on syntactic processing in a nonhuman primate. *Science* 303:377–80.

Fraser, O. N., and T. Bugnyar. 2010. Do ravens show consolation? Responses to distressed others. PLoS ONE 5:1–8. e10605.

Fraser, O. N., and T. Bugnyar. 2011. Ravens reconcile after aggressive conflicts with valuable partners. PLoS ONE 6:1–5. e18118.

Freed, P. J., T. K. Yanagihara, J. Hirsch, and J. J. Mann. 2009. Neural mechanisms of grief regulation. *Biological Pyschiatry* 66:33–40.

Frings, H., M. Frings, J. Jumber, R-G. Busnel, J. Giban, and P. Gramet. 1958. Reactions of American and French species of *Corvus* and *Larus* to recorded communication signals tested reciprocally. *Ecology* 39:126–31.

Gale, S. D., and D. J. Perkel. 2006. Physiological properties of zebra finch ventral tegmental area and substantia nigra pars compacta neurons. *Journal of Neurophysiology* 96:2295–2306.

Gale, S. D., and D. J. Perkel. 2010. A basal ganglia pathway drives selective auditory responses in songbird dopaminergic neurons via disinhibition. *Journal of Neuroscience* 30:1027–37.

Garamszegi, L. Z., M. Eens, D. Z. Pavlova, J. M. Avilés, and A. P. Møller. 2007. A comparative study of the function of heterospecific vocal mimicry in European passerines. *Behavioral Ecology* 18:1001–9.

Gehring, T. M. 1993. Potential predatory attack by Common Ravens on porcupines. *Wilson Bulletin* 105:524–25.

Gilbert, B. 1988. Goodbye, hello. *Sports Illustrated* 69:108–22.

Gill, F. B. 2007. *Ornithology,* 3rd ed. New York, NY: W. H. Freeman and Company.

Giret, N., F. Péron, J. Lindová, L. Tichotová, L. Nagle, M. Kreutzer, F. Tymr, and D. Bovet. 2010. Referential learning of French and Czech labels in African grey parrots (*Psittacus erithacus*): Different methods yield contrasting results. *Behavioural Processes* 85:90–98.

Glees, P. 1952. Ludwig Edinger 1855–1918. *Journal of Neurophysiology* 15:251–55.

Gobbini, M. I., and J. V Haxby. 2007. Neural systems for recognition of familiar faces. *Neuropsychologia* 45:32–41.

Goldin, P. R., C. A. C. Hutcherson, K. N. Ochsner, G. H. Glover, J. D. E. Gabrieli, and J. J. Gross. 2005. The neural bases of amusement and sadness: a comparison of block contrast and subject-specific emotion intensity regression approaches. *NeuroImage* 27:26–36.

Goller, F., and O. N. Larsen. 1997. A new mechanism of sound generation in songbirds. *Proceedings of the National Academy of Science USA* 94:14787–91.

Goodson, J. L. 2005. The vertebrate social behavior network: Evolutionary themes and variations. *Hormones and Behavior* 48:11–22.

Goodson, J. L., and D. Kabelik. 2009. Dynamic limbic networks and social diversity in vertebrates: From neural context to neuromodulatory patterning. *Frontiers in Neuroendocrinology* 30:429–41.

Goodson, J. L., D. Kabelik, A. M. Kelly, J. Rinaldi, and J. D. Klatt. 2009. Midbrain dopamine neurons reflect affiliation phenotypes in finches and are tightly coupled to courtship. *Proceedings of the National Academy of Science USA* 106:8737–42.

Goodwin, D. 1986. *Crows of the World,* 2nd ed. Seattle, WA: University of Washington Press.

Gorenzel, W. P., and T. P. Salmon. 1993. Tape-recorded calls disperse American crows from urban roosts. *Wildlife Society Bulletin* 21:334–38.

Grace, A. A., S. B. Floresco, Y. Goto, and D. J. Lodge. 2007. Regulation of firing of dopaminergic neurons and control of goal-directed behaviors. *Trends in Neurosciences* 30:220–27.

Griesser, M. 2008. Referential calls signal predator behavior in a group-living bird species. *Current Biology* 18:69–73.

Griffin, A. S. 2004. Social learning about predators: A review and prospectus. *Learning & Behavior* 32:131–40.

Griffin, A.S., and H.M. Boyce. 2009. Indian mynahs, *Acridotheres tristis,* learn about dangerous places by observing the fate of others. *Animal Behaviour* 78:79–84.

Griver, J. 2004. *Curio, a Shetland Sheepdog, Meets the Crow*. Pacific Palisades, CA: Compsych Systems, Inc.

Güntürkün, O. 2005. The avian "prefrontal cortex" and cognition. *Current Opinion in Neurobiology* 15:686–93.

Hailman, J. P. 1990. Blue jay mimics osprey. *Florida Field Naturalist* 18:81–82.

Hartmann, B., and O. Güntürkün. 1998. Selective deficits in reversal learning after neostriatum caudolaterale lesions in pigeons: Possible behavioral equivalencies to the mammalian prefrontal system. *Behavioural Brain Research* 96:125–33.

Hauser, M. D. 2000. *Wild Minds*. New York, NY: Henry Holt and Company.

Hauser, M. D., and C. Caffrey. 1994. Anti-predator response to raptor calls in wild crows, *Corvus brachrhynchos hesperis*. *Animal Behaviour* 48:1469–71.

Healy, S. D., and K. R. Krebs. 1993. Development of hippocampal specialization in a food-storing bird. *Behavioural Brain Research* 53:127–31.

Heinrich, B. 1995. An experimental investigation of insight in common ravens (*Corvus corax*). *Auk* 112:994–1003.

Heinrich, B. 1999. *Mind of the Raven*. New York, NY: Harper Collins Publishers.

Heinrich, B. 2000. Testing insight in ravens. Pages 289–305. In *The Evolution of Cognition,* edited by L. Huber and C. M. Heyes. Cambridge, MA: MIT Press.

Heinrich, B. 2010. *The Nesting Season: Cuckoos, Cuckolds, and the Invention of Monogamy.* Cambridge, MA: Belknap Press.

Heinrich, B., and J. M. Marzluff. 1991. Do common ravens yell because they want to attract others? *Behavioral Ecology and Sociobiology* 28:13–21.

Heinrich, B., and J. W. Pepper. 1998. Influence of competitors on caching behaviour in the common raven, *Corvus corax*. *Animal Behaviour* 56:1083–90.

Heinrich, B., and R. Smolker. 1998. Play in common ravens (*Corvus corax*). Pages 27–44. In *Animal Play: Evolutionary, Comparative, and Ecological Perspectives,* edited by M. Bekoff and J. A. Byers. Cambridge, UK: Cambridge University Press.

Heyers, D., M. Zapka, M. Hoffmeister, J. M. Wild, and H. Mouritsen. 2010. Magnetic field changes activate the trigeminal brain-stem complex in a migratory bird. *Proceedings of the National Academy of Science USA* 107:9394–99.

Heyers, D., M. Manns, H. Luksch, O. Güntürkün, and H. Mouritsen. 2007. A visual pathway links brain structures active during magnetic compass orientation in migratory birds. PLoS One 9:e937.

Heyes, C. M. 1994. Social learning in animals: Categories and mechanisms. *Biological Review* 69:207–31.

Heyes, C. M. 1998. Theory of mind in nonhuman primates. *Behavioral and Brain Sciences* 21:101–48.

Higuchi, H., and R. Kurosawa. 2010. *Karasu no shizenshi—keitou kara asobikoudou made* (A natural history of crows—from phylogeny to play behavior). Hokkaido, Japan: Hokkaidou Daigaku Shuppankai (Hokkaido University Press).

Hildreth, A. M., S. M. Vantassel, and S. E. Hygnstrom. 2010. Feral cats and their management. University of Nebraska–Lincoln Extension publication EC1781, http://extension.unl.edu/publications.

Hoeschele. J. P. 1989. A new look at the uncommon common crow. *The Conservationist* (State of New York, Department of Environmental Conservation) March–April 1989:16–19.

Holden, C. 2004. The origin of speech. *Science* 303:1316–19.

Hope, S. 1980. Call form in relation to function in the Steller's jay. *The American Naturalist* 116:788–820.

Horn, G. 1981. Neural mechanisms of learning: An analysis of imprinting in the domestic chick. *Proceedings of the Royal Society of London B* 213:101–37.

Horn, G., and B. J. McCabe. 1984. Predispositions and preferences: Effects on imprinting of lesions to the chick brain. *Animal Behaviour* 32:288–92.

Horn, G., B. J. McCabe, and J. Cipolla-Neto. 1983. Imprinting in the domestic chick: The role of each side of the hyperstriatum ventrale in acquisition and retention. *Experimental Brain Research* 53:91–98.

Hornfeld, S. H., J. Terkel, and A. Barnea. 2010. Neurons recruited in the nidopallium caudale, following changes in social environment, derive from the same original population. *Behavioural Brain Research* 208:643–45.

Hunt, G. R., M. C. Corballis, and R. D. Gray. 2001. Laterality in tool manufacture by crows. *Nature* 414:707.

Iacoboni, M. 2005. Neural mechanisms of imitation. *Current Opinion in Neurobiology* 15:632–37.

Insel, T. R., and R. D. Fernald. 2004. How the brain processes social information: Searching for the social brain. *Annual Review of Neuroscience* 27:697–722.

Iordanova, M. D., G. P. McNally, and R. F. Westbrook. 2006. Opioid receptors in the nucleus accumbens regulate attentional learning in the blocking paradigm. *Journal of Neuroscience* 26:4036–45.

Iwaniuk, A. N., and J. E. Nelson. 2003. Developmental differences are correlated with relative brain size in birds: A comparative analysis. *Canadian Journal of Zoology* 81:1913–28.

Izawa, E-I., T. Kusayama, and S. Watanabe. 2005. Foot-use laterality in the Japanese jungle crow (*Corvus macrorhynchos*). *Behavioural Processes* 69:357–62.

Jarvis, E. D. 2004. Learned birdsong and the neurobiology of human language. *Annals of the New York Academy of Sciences* 1016:749–77.

Jepsen, G. L. 1966. Early Eocene bat from Wyoming. *Science* 154:1333–39.

Jitsumori, M., and H. Makino. 2004. Recognition of static and dynamic images of depth-rotated human faces by pigeons. *Learning & Behavior,* 32, 145–56.

Jurkevich, A., R. Grossmann, J. Balthazart, and C. Viglietti-Panzica. 2001. Gender-related changes in the avian vasotocin system during ontogeny. *Microscopy Research and Technique* 55:27–36.

Kalenscher, T., S. Windmann, B. Diekamp, J. Rose, O. Güntürkün, and M. Colombo. 2005. Single units in the pigeon brain integrate reward amount and time-to-reward in an impulsive choice task. *Current Biology* 15:594–602.

Kamil, A. C., R. P. Balda, and S. Good. 1999. Patterns of movement and orientation during caching and recovery by Clark's nutcrackers (*Nucifraga columbiana*). *Animal Behaviour* 57:1327–35.

Kandel, E. R. 2009. The biology of memory: A forty-year perspective. *Journal of Neuroscience* 29:12748–56.

Kaplan, G. 2004. Meaningful communication in primates, birds, and other animals. Pages 189–223. In *Comparative Vertebrate Cognition: Are Primates Superior to Non-Primates?,* edited by L. J. Rogers and G. Kaplan. New York, NY: Kluwer Academic Press.

Kaplan, G. 2008. Alarm calls and referentiality in Australian magpies: Between midbrain and forebrain, can a case be made for complex cognition? *Brain Research Bulletin* 76:253–63.

Kaster, R. A. 2011. *Saturnalia*. Translation of the text by Ambrosius Aurelius Theodosius Macrobius. Cambridge, MA: Harvard University Press.

Kavaliers, M., D. D. Colwell, and E. Choleris. 2003. Learning to fear and cope with a natural stressor: Individuality and socially acquired corticosterone and avoidance responses to biting flies. *Hormones and Behavior* 43:99–107.

Kelley, L. A., R. L. Coe, J. R. Madden, and S. D. Healy. 2008. Vocal mimicry in songbirds. *Animal Behaviour* 76:521–28.

Kendrick, K. M., and B. A. Baldwin. 1987. Cells in temporal cortex of conscious sheep can respond preferentially to the sight of faces. *Science* 236:448–50.

Kilham, L. 1982. Common crows pulling the tail and stealing food from a river otter. *Florida Field Naturalist* 10:39–40.

Kilham, L. 1985. Some breeding season vocalizations of American crows in Florida. *Florida Field Naturalist* 13:49–76.

Kilham, L. 1986. Vocalizations by female American crows early in the nesting period. *Journal of Field Ornithology* 57:309–10.

Kilham, L. 1989. *The American Crow and the Common Raven*. College Station, TX: Texas A&M University Press.

Kim, H., S. Shimojo, and J. P. O'Doherty. 2006. Is avoiding an aversive outcome rewarding? The neural substrates of avoidance learning in the human brain. PLoS Biology 4:1453–61, e233.

King, B. J. 2010. *Being with Animals: Why We Are Obsessed with the Furry, Scaly, Feathered Creatures Who Populate Our World.* New York, NY: Doubleday.

Kirsch, J. A., I. Vlachos, M. Hausmann, J. Rose, M. Y. Yim, A. Aertsen, and O. Güntürkün. 2009. Neuronal encoding of meaning: Establishing category-selective response patterns in the avian "prefrontal cortex." *Behavioural Brain Research* 198:214–23.

Knörnschild, M., M. Nagy, M. Metz, F. Mayer, and O. von Helversen. 2010. Complex vocal imitation during ontogeny in a bat. *Biology Letters* 6:156–59.

Köppl, C., G. A. Manley, and M. Konishi. 2000. Auditory processing in birds. *Current Opinion in Neurobiology* 10:474–81.

Korte, S. M. 2001. Corticosteroids in relation to fear, anxiety and psychopathology. *Neuroscience and Biobehavioral Reviews* 25:117–42.

Kotaleski, J. H., and K. T. Blackwell. 2010. Modelling the molecular mechanisms of synaptic plasticity using systems biology approaches. *Nature Reviews Neuroscience* 11:239–51.

Kotegawa, T., T. Abe, and K. Tsutsui. 1997. Inhibitory role of opioid peptides in the regulation of aggressive and sexual behaviors in male Japanese quails. *Journal of Experimental Zoology* 277:146–54.

Krebs, J. R., N. S. Clayton, R. R. Hampton, and S. J. Shettleworth. 1995. Effects of photoperiod on food-storing and the hippocampus in birds. *Neuroreport* 6:1701–4.

Kretzschmar, C., T. Kalenscher, O. Güntürkün, and C. Kaernbach. 2008. Echoic memory in pigeons. *Behavioural Processes* 79:105–10.

Kubikova, L., K. Wada, and E. D. Jarvis. 2009. Dopamine receptors in a songbird brain. *Journal of Comparative Neurology* 518:741–69.

Langen, T. A. 1996a. Social learning of a novel foraging skill by white-throated magpie-jays (*Calocitta formosa,* Corvidae): A field experiment. *Ethology* 102:157–66.

Langen, T. A. 1996b. Skill acquisition and the timing of natal dispersal in the white-throated magpie-jay, *Calocitta formosa*. *Animal Behaviour* 51:575–88.

Lavenex, P. B. 2000. Lesions in the budgerigar vocal control nucleus NLC affect production, but not memory, of English words and natural vocalizations. Journal of Comparative Neurology 421:437–60.

Laverghetta, A. V., and T. Shimizu. 2003. Organization of the ectostriatum based on afferent connections in the zebra finch (*Taeniopygia guttata*). *Brain Research* 963:101–12.

LeDoux, J. E. 2000. Emotion circuits in the brain. *Annual Review of Neuroscience* 23:155–84.

LeDoux, J. E., and E. A. Phelps. 2008. Emotional networks in the brain. Pages 159–79. In *Handbook of Emotions,* 3rd ed., edited by M. Lewis, J. M. Haviland-Jones, and L. F. Barrett. New York, NY: Guilford Press.

Lee, H., and J. J. Kim. 1998. Amygdalar NMDA receptors are critical for new fear learning in previously fear-conditioned rats. *Journal of Neuroscience* 18:8444–54.

Lefebvre, L. 1995. The opening of milk bottles by birds: Evidence for accelerating learning rates, but against the wave-of-advance model of cultural transmission. *Behavioural Processes* 34: 43–53.

Lefebvre, L., and D. Sol. 2008. Brains, lifestyles and cognition: Are there general trends? *Brain, Behavior and Evolution* 72:135–44.

Lefebvre, L., S. M. Reader, and D. Sol. 2004. Brains, innovations and evolution in birds and primates. *Brain Behavior and Evolution* 63:233–46.

Lefebvre, L., P. Whittle, E. Lascaris, and A. Finkelstein. 1997. Feeding innovations and forebrain size in birds. *Animal Behaviour* 53:549–60.

Libet, B. 2004. *Mind Time*. Cambridge, MA: Harvard University Press.

Linden, D. J. 2007. *The Accidental Mind*. Cambridge, MA: Harvard University Press.

Lissek, S., and O. Güntürkün. 2003. Dissociation of extinction and behavioral disinhibition: The role of NMDA receptors in the pigeon associative forebrain during extinction. *Journal of Neuroscience* 23:8119–24.

Llinás, R. R., and M. Steriade. 2006. Bursting of thalamic neurons and states of vigilance. *Journal of Neurophysiology* 95:3297–3308.

Lord, K., M. Feinstein, and R. Coppinger. 2009. Barking and mobbing. *Behavioural Processes* 81:358–68.

Lorenz, K. Z. 1937. The companion in the bird's world. *Auk* 54:245–73.

Lorenz, K. Z. 1952. *King Solomon's Ring*. New York, NY: Thomas Y. Crowell.

Lorenz, K. Z. 1981. *The Foundations of Ethology*. Touchstone, New York.

Lutz, B. 2009. Endocannabinoid signals in the control of emotion. *Current Opinion in Pharmacology* 9:46–52.

Maguire, E. A., N. Burgess, J. G. Donnett, R. S. J. Frackowiak, C. D. Firth, and J. O'Keefe. 1998. Knowing where and getting there: A human navigation network. *Science* 280:921–24.

Mallory, F. F. 1968. An ingenious hunting behavior in the common raven (*Corvus corax*). *Ontario Field Biologist* 31:53.

Maren, S. 2005. Central and basolateral amygdala neurons crash the aversive conditioning party: Theoretical comment on Rorick-Kehn and Steinmetz (2005). *Behavioral Neuroscience* 119:1406–10.

Marley, E., and T. J. Seller. 1974. Effects of nicotine given into the brain of fowls. *British Journal of Pharmacology* 51:335–46.

Martin, J. T., N. Delanerolle, and R. E. Phillips. 1979. Avian archistriatal control of fear-motivated behavior and adrenocortical function. *Behavioural Processes* 4:283–93.

Marzluff, J. M., and R. P. Balda. 1988a. Pairing patterns and fitness in a free-ranging population of pinyon jays: What do they reveal about mate choice? *Condor* 90:201–13.

Marzluff, J. M., and R. P. Balda. 1988b. The advantages of, and constraints forcing, mate fidelity in pinyon jays. *Auk* 105:286–95.

Marzluff, J. M., and R. P. Balda. 1992. *The Pinyon Jay*. London: Poyser.

Marzluff, J. M., B. Heinrich, and C. S. Marzluff. 1996. Communal roosts of common ravens are mobile information centers. *Animal Behaviour*. 51:89–103.

Marzluff, J. M., and T. Angell. 2005a. *In the Company of Crows and Ravens*. New Haven, CT: Yale University Press.

Marzluff, J. M., and T. Angell. 2005b. Cultural coevolution: How the human bond with crows and ravens extends theory and raises new questions. *Journal of Ecological Anthropology* 9:67–73.

Marzluff, J. M., and C. Marzluff. 2011. *Dog Days, Raven Nights*. New Haven, CT: Yale University Press.

Marzluff, J. M., J. Walls, H. N. Cornell, J. C. Withey, and D. P. Craig. 2010. Lasting recognition of threatening people by wild American crows. *Animal Behaviour* 79:699–707.

McGaugh, J. L. 2004. The amygdala modulates the consolidation of memories of emotionally arousing experiences. *Annual Review of Neuroscience* 27:1–28.

Mech, D. 1970. *The Wolf*. Minneapolis, MN: University of Minnesota Press.

Mee, A., B. A. Rideout, J. A. Hamber, J. N. Todd, G. Austin, M. Clark, and M. P. Wallace. 2007. Junk ingestion and nestling mortality in a reintroduced population of California condors *Gymnogyps californicus*. *Bird Conservation International* 17:119–30.

Mehlhorn, J., G. R. Hunt, R. D. Gray, G. Rehkämper, and O. Güntürkün. 2010. Tool-making New Caledonian crows have large associative brain areas. *Brain, Behavior and Evolution* 75:63–70.

Milad, M. R., and G. J. Quirk. 2002. Neurons in medial prefrontal cortex signal memory for fear extinction. *Nature* 420:70–74.

Moffett, A. T. 1984. Ravens sliding in snow. *British Birds* 77:321–22.

Moles, A., B. L. Kieffer, and F. R. D'Amato. 2004. Deficit in attachment behavior in mice lacking the µ-opioid receptor gene. *Science* 304:1983–86.

Montagnese, C. M., S. E. Mezey, and A. Csillag. 2003. Efferent connections of the dorsomedial thalamic nuclei of the domestic chick (*Gallus domesticus*). *Journal of Comparative Neurology* 459:301–26.

Montevecchi, W. A. 1978. Corvids using objects to displace gulls from nests. *Condor* 80:349.

Mouritsen, H., G. Feenders, M. Liedvogel, K. Wada, and E. D. Jarvis. 2005. Night-vision brain area in migratory songbirds. *Proceedings of the National Academy of Science USA* 102:8339–44.

Nardi, D., and V. P. Bingman. 2007. Asymmetrical participation of the left and right hippocampus for representing environmental geometry in homing pigeons. *Behavioural Brain Research* 178:160–71.

Nicolaus, L. K., and J. F. Cassel. 1983. Taste-aversion conditioning of crows to control predation on eggs. *Science* 220:212–214.

Nihei, Y., and H. Higuchi. 2001. When and where did crows learn to use automobiles as nutcrackers? *Tohoku Psychologica Folia* 60:93–97.

Nomura, T., H. Mitsuharu, and O. Noriko. 2009. Reelin radial fibers and cortical evolution: Insights from comparative analysis of the mammalian and avian telencephalon. *Development Growth & Differentiation* 51:287–297.

Nottebohm, F. 1984. Birdsong as a model in which to study brain processes related to learning. *Condor* 86:227–36.

O'Connor, M. F., M. R. Irwin, and D. K. Wellisch. 2009. When grief heats up: Pro-inflammatory cytokines predict regional brain activation. *NeuroImage* 47:891–96.

Öhman, A. 2008. Fear and anxiety, overlaps and dissociations. Pages 709–29. In *Handbook of Emotions,* 3rd ed., edited by M. Lewis, J. M. Haviland-Jones, and L. F. Barrett. New York, NY: Guilford Press.

Okanoya, K., M. Ikebuchi, H. Uno, and S. Watanabe. 2001. Left-side dominance for song discrimination in Bengalese finches *(Lonchura striata var. domestica). Animal Cognition* 4:241–45.

Olsson, A., and E. A. Phelps. 2007. Social learning of fear. *Nature Neuroscience* 10:1095–1102.

Olsson, A., K. I. Nearing, and E. A. Phelps. 2007. Learning fears by observing others: The neural systems of social fear transmission. *Social Cognitive and Affective Neuroscience* 2:3–11.

Ortega, L. J., K. Stoppa, O. Güntürkün, and N. F. Troje. 2008. Limits of intraocular and interocular transfer in pigeons. *Behavioural Brain Research* 193:69–78.

Palermo, R., and G. Rhodes. 2007. Are you always on my mind? A review of how face perception and attention interact. *Neuropyschologia* 45:75–92.

Panksepp, J., B. H. Herman, T. Vilberg, P. Bishop, and F. G. DeEskinazi. 1978. Endogenous opioids and social behavior. *Neuroscience & Biobehavioral Reviews* 4:473–87.

Panksepp, J., S. Siviy, and L. Normansell. 1984. The psychobiology of play: Theoretical and methodological perspectives. *Neuroscience & Biobehavioral Reviews* 8:465–92.

Parr, L. A., T. Dove, and W. D. Hopkins. 1998. Why faces may be special: Evidence of the inversion effect in chimpanzees. *Journal of Cognitive Neuroscience* 10:615–22.

Pas-y-Mino, C. G., A. B. Bond, A. C. Kamil, and R. P. Balda. 2004. Pinyon jays use transitive inference to predict social dominance. *Nature* 430:778–81.

Patel, S. N., N. S. Clayton, and J. R. Krebs. 1997. Spatial learning induces neurogenesis in the avian brain. *Behavioural Brain Research* 89:115–28.

Pepperberg, I. M. 2007. Grey parrots do not always "parrot": The roles of imitation and phonological awareness in the creation of new labels from existing vocalizations. *Language Sciences* 29:1–13.

Phelps, E. A. 2006. Emotion and cognition: Insights from studies of the human amygdala. *Annual Review of Psychology* 57:27–53.

Phillmore, L. S. 2008. Discrimination: From behaviour to brain. *Behavioural Processes* 77:285–97.

Piazza, P. V., and M. Le Moal. 1997. Glucocorticoids as a biological substrate of reward: Physiological and pathophysiological implications. *Brain Research Reviews* 25:359–72.

Pollok, B., H. Prior, and O. Güntürkün. 2000. Development of object permanence in food-storing magpies (*Pica pica*). *Journal of Comparative Psychology* 114:148–57.

Poole, J. H., P. L. Tyack, A. S. Stoeger-Horwath, and S. Watwood. 2005. Animal behaviour: Elephants are capable of vocal learning. *Nature* 434: 455–56.

Prather, J. F., S. Peters, S. Nowicki, and R. Mooney. 2008. Precise auditory-vocal mirroring in neurons for learned vocal communication. *Nature* 451: 305–10.

Premack, D. 2004. Is language the key to human intelligence? *Science* 303:318–20.

Prior, H. A. Schwarz, and O. Güntürkün. 2008. Mirror-induced behavior in the magpie *(Pica pica)*: Evidence of self-recognition. *PLoS Biology* 6:1–9, e202.

Raby, C. R., D. M. Alexis, A. Dickinson, and N. S. Clayton. 2007. Planning for the future by western scrub-jays. *Nature* 445:919–21.

Ratcliffe, D. 1997. *The Raven*. London: Poyser.

Ratnayake, C. P., E. Goodale, and S. W. Kotagama. 2010. Two sympatric species of passerine birds imitate the same raptor calls in alarm contexts. *Naturwissenschaften* 97:103–8.

Rattenborg, N. C., D. Martinez-Gonzalez, and J. A. Lesku. 2009. Avian sleep homeostasis: Convergent evolution of complex brains, cognition and sleep functions in mammals and birds. *Neuroscience and Biobehavioural Reviews* 33:253–70.

Reiner, A., K. Yamamoto, and H. J. Karten. 2005. Organization and evolution of the avian forebrain. *Anatomical Record Part A* 287A:1080–1102.

Reiner, A., D. J. Perkel, L. L. Bruce, A. B. Butler, A. Csillag, W. Kuenzel, L. Medina, G. Paxinos, T. Shimizu, G. Striedter, M. Wild, G. F. Ball, S. Durand, O. Gütürkün, D. W. Lee, C. V. Mello, A. Powers, S. A. White, G. Hough, L. Kubikova, T. V. Smulders, K. Wada, J. Dugas-Ford, S. Husband, K. Yamamoto, J. Yu, C. Siang, and E. D. Jarvis. 2004. Revised nomenclature for avian telencephalon and some related brain-stem nuclei. *Journal of Comparative Neurology* 473:377–414.

Reinhardt, I., A. Jansen, T. Kellermann, A. Schüppen, N. Kohn, A. L. Gerlach, and T. Kircher. 2010. Neural correlates of aversive conditioning: Development of a functional imaging paradigm for the investigation of anxiety disorders. *European Archives in Psychiatry and Clinical Neuroscience* 260:443–53.

Richards, D. B., and N. S. Thompson. 1978. Critical properties of the assembly call of the common American crow. *Behaviour* 64:184–204.

Ricklefs, R. E. 2004. The cognitive face of avian life histories. *Wilson Bulletin* 116:119–33.

Riters, L. V., and G. F. Ball. 1999. Lesions to the medial preoptic area affect singing in the male European starling (*Sturnus vulgaris*). *Hormones and Behavior* 36:276–86.

Riters, L. V., P. Absil, and J. Balthazart. 1998. Effects of brain testosterone implants on appetitive and consummatory components of male sexual behavior in Japanese quail. *Brain Research Bulletin* 47:69–79.

Ritz, T., P. Thalau, J. B. Phillips, R. Wiltschko, and W. Wiltschko. 2004. Resonance effects indicate a radical-pair mechanism for avian magnetic compass. *Nature* 429:177–80.

Rizzolatti, G., and L. Craighero. 2004. The mirror-neuron system. *Annual Review of Neuroscience* 27:169–92.

Rizzolatti, G., and M. Fabbri-Destro. 2008. The mirror system and its role in social cognition. *Current Opinion in Neurobiology* 18:179–84.

Rogers, L. J. 2004. Increasing the brain's capacity: Neocortex, new neurons, and hemispheric specialization. Pages 289–323. In *Comparative Vertebrate Cognition: Are Primates Superior to Non-Primates?*, edited by L. J. Rogers and G. Kaplan. New York, NY: Kluwer Academic Press.

Rolls, E. T. 2007. The representation of information about faces in the temporal and frontal lobes. *Neuropsychologia* 45:124–43.

Rose, J., and M. Colombo. 2005. Neural correlates of executive control in the avian brain. *PLoS Biology* 3:e190.

Rose, S. P., and M. G. Stewart. 1999. Cellular correlates of stages of memory formation in the chick following passive avoidance training. *Behavioural Brain Research* 98:237–43.

Rowe, C., and J. Skelhorn. 2004. Avian psychology and communication. *Proceedings of the Royal Society of London B* 271:1435–42.

Rowley, I. 1969. An evaluation of predation by "crows" on young lambs. *CSIRO Wildlife Research* 14:153–79.

Saint-Dizier, H., P. Constantin, D. C. Davies, C. Leterrier, F. Lévy, and S. Richard. 2009. Subdivisions of the arcopallium/posterior pallial amygdala complex are differentially involved in the control of fear behaviour in the Japanese quail. *Brain Research Bulletin* 79:288–95.

Salamone, J. D. 2007. Functions of mesolimbic dopamine: Changing concepts and shifting paradigms. *Psychopharmacology* 191:389.

Salamone, J. D., M. Correa, A. Farrar, and S. M. Mingote. 2007. Effort-related functions of nucleus accumbens dopamine and associated forebrain circuits. *Psychopharmacology* 191:461–82.

Salzen, E. A., A. J. Williamson, and D. M. Parker. 1979. The effects of forebrain lesions on innate and imprinted colour, brightness and shape preferences in domestic chicks. *Behavioural Processes* 4:295–313.

Samson, R. D., and D. Paré. 2005. Activity-dependent synaptic plasticity in the central nucleus of the amygdala. *Journal of Neuroscience* 25:1847–55.

Santini, E., R. U. Muller, and G. J. Quirk. 2001. Consolidation of extinction learning involves transfer from NMDA-independent to NMDA-dependent memory. *Journal of Neuroscience* 21:9009–17.

Sax, B. 2011. *City of Ravens*. London: Duckworth.

Scharff, C., and S. Haesler. 2005. An evolutionary perspective on FoxP2: Strictly for the birds? *Current Opinion in Neurobiology* 15:684–703.

Schloegl, C., K. Kotrschal, and T. Bugnyar. 2007. Gaze following in common ravens, *Corvus corax*: Ontogeny and habituation. *Animal Behaviour* 74:769–78.

Schultz, W. 2006. Behavioral theories and the neurophysiology of reward. *Annual Review of Psychology* 57:87–115.

Schultz, W. 2007. Behavioral dopamine signals. *Trends in Neurosciences* 30:203–10.

Seed, A. M., N. S. Clayton, and N. J. Emery. 2007. Postconflict third-party affiliation in rooks, *Corvus frugilegus*. *Current Biology* 17:152–58.

Seed, A. M., N. S. Clayton, and N. J. Emery. 2008. Cooperative problem solving in rooks *(Corvus frugilegus)*. *Proceeding of the Royal Society B* 275:1421–29.

Seller, T. J. 1981. Midbrain vocalization centres in birds. *Trends in Neurosciences* 12:301–3.

Shaw, B. K. 2000. Involvement of a midbrain vocal nucleus in the production of both the acoustic and postural components of crowing behavior in Japanese quail. *Journal of Comparative Physiology A* 186:747–57.

Sherry, D. F., and J. S. Hoshooley. 2009. The seasonal hippocampus of food-storing birds. *Behavioural Processes* 80:334–38.

Sherry, D. F., and J. B. Mitchell. 2007. Neuroethology of foraging. Pages 61–102. In *Foraging, Behavior and Ecology,* edited by D. W. Stephens, J. S. Brown, and R. C. Ydenberg. Chicago, IL: University of Chicago Press.

Simpson, H. B., and D. S. Vicario. 1990. Brain pathways for learned and unlearned vocalizations differ in zebra finches. *Journal of Neuroscience* 10:1541–56.

Siviy, S. M. 1998. Neurobiological substrates of play behavior: Glimpses into the structure and function of mammalian playfulness. Pages 221–42. In *Animal Play: Evolutionary, Comparative, and Ecological Perspectives,* edited by M. Bekoff and J. A. Byers. Cambridge, UK: Cambridge University Press.

Siviy, S. M., K. A. Harrison, and I. S. McGregor. 2006. Fear, risk assessment, and playfulness in the juvenile rat. *Behavioral Neuroscience* 120:49–59.

Soderstrom, K., and F. Johnson. 2000. CB1 cannabinoid receptor expression in brain regions associated with zebra finch song control. *Brain Research* 857:151–57.

Sol, D., R. P. Duncan, T. M. Blackburn, P. Cassey, and L .Lefebvre. 2005. Big brains, enhanced cognition, and response of birds to novel environments. *Proceedings of the National Academy of Science USA* 102:5460–65.

Sol, D., T. Székely, A. Liker, and L. Lefebvre. 2007. Big-brained birds survive better in nature. *Proceedings of the Royal Society of London B* 274:763–69.

Spear, L. 1988. The Halloween mask episode: A gull researcher learns the barefaced facts about western gulls. *Natural History*. June 1988.

Stahnisch, F. W. 2008. Ludwig Edinger (1855–1918). *Journal of Neurology* 255:147–48

Stewart, M. G., P. Kabai, E. Harrison, R. J. Steele, M. Kossut, M. Gierdalski, and A. Csillag. 1996. The involvement of dopamine in the striatum in passive avoidance training in the chick. *Neuroscience* 70:7–14.

Straub, A. 2007. An intelligent crow beats a lab. *Science* 316:688.

Striedter, G. F. 2005. *Principles of Brain Evolution*. Sunderland, MA: Sinauer Associates.

Sugai, R., S. Azami, H. Shiga, T. Watanabe, H. Sadamoto, S. Kobayashi, D. Hatakeyama, Y. Fujito, K. Lukowiak, and E. Ito. 2007. One-trial conditioned taste aversion in Lymnaea: Good and poor performers in long-term memory acquisition. *Journal of Experimental Biology* 210:1225–37.

Surman, M. 2005. Liver-lovin' crows causing toads to pop? Vet says so. *Seattle Times,* April 29, 2005.

Tate, A. J., H. Fischer, A. E. Leigh, and K. M. Kendrick. 2006. Behavioural and neurophysiological evidence for face identity and face emotion processing in animals. *Philosophical Transactions of the Royal Society of London B* 361:2155–72.

Taylor, A. H., F. S. Medina, J. C. Holzhaider, L. J. Hearne, G. R. Hunt, and R. D. Gray. 2010a. An investigation into the cognition behind spontaneous string pulling in New Caledonian crows. *PloS ONE* 5: e9345.

Taylor, A. H., D. Elliffe, G. R. Hunt, and R. D. Gray. 2010b. Complex cognition and behavioural innovation in New Caledonian crows. *Proceedings of the Royal Society of London B* 277:2637–43.

Theunissen, F. E., and S. S. Shaevitz. 2006. Auditory processing of vocal sounds in birds. *Current Opinion in Neurobiology* 16:400–407.

Thompson, N. S. 1968. Counting and communication in crows. *Communications in Behavioral Biology A* 2:223–25.

Thompson, N. S. 1982. A comparison of cawing in the European carrion crow *(Corvus corone)* and the American common crow (*Corvus brachyrhynchos*). *Behaviour* 80:106–17.

Thorpe, W. H. 1966. Ritualization in ontogeny: I. Animal play. *Philosophical Transactions of the Royal Society of London B* 251:311–19, 351–58.

Thorpe, W. H., and M. E. W. North. 1966. Vocal imitation in the tropical boubou shrike *Laniarius aethiopicus* major as a means of establishing and maintaining social bonds. *Ibis* 108:432–35.

Tramontin, A. D., and E. A. Brenowitz. 2000. Seasonal plasticity in the adult brain. *Trends in Neurosciences* 23:251–58.

Trezza, V., and L. J. Vanderschuren. 2008. Bidirectional cannabinoid modulation of social behavior in adolescent rats. *Psychopharmacology* 197:217–27.

Trezza, V., P. J. Baarendse, and L. J. Vanderschuren. 2010. The pleasures of play: Pharmacological insights into social reward mechanisms. *Trends in Pharmacological Sciences* 31:463–69.

Vallortigara, G., and R. J. Andrew. 1994. Differential involvement of right and left hemisphere in individual recognition in the domestic chick. *Behavioural Processes* 33:41–58.

Van der Linden, A., V. Van Meir, T. Boumans, C. Poirier, and J. Blathazart. 2009. MRI in small brains displaying extensive plasticity. *Trends in Neurosciences* 32:257–66.

van Hierden, Y. M., S. M. Korte, E. W. Ruesink, C. G. van Reenen, C. B. Engle, G. A. Korte-Bouws, J. M. Koolhaas, and H. J. Blokhuis. 2002. Adrenocortical reactivity and central serotonin and dopamine turnover in young chicks from a high and low feather-pecking line of laying hens. *Physiology & Behavior* 75:653–59.

Vanderschuren, L. J., E. A. Stein, V. M. Wiegant, and J. M. Van Ree. 1995. Social play alters regional brain opioid receptor binding in juvenile rats. *Brain Research* 680:148–56.

Vanderschuren, L. J., V. Trezza, S. Griffioen-Roose, O. J. G. Schiepers, N. Van Leeuwen, T. J. De Vries, and A. N. Schoffelmeer. 2008. Methylphenidate disrupts social play behavior in adolescent rats. *Neuropsychopharmacology* 33:2946–56.

Velella, R. 2009. Charles Dickens with his raven. The Edgar A. Poe Bicentennial Desk Calendar. poecalendar blogspot.com. February 7, 2009.

Vuilleumier, P., and G. Pourtois. 2007. Distributed and interactive brain mechanisms during emotion face perception: Evidence from functional neuroimaging. *Neuropsychologia* 45:174–94.

Wager, T. D., L. F. Barrett, E. Bliss-Moreau, K. A. Lindquist, S. Duncan, H. Kober, J. Joseph, M. Davidson, and J. Mize. 2008. The neuroimaging of emotion. Pages 249–67. In *Handbook of Emotions,* 3rd ed., edited by M. Lewis, J. M. Haviland-Jones, and L. F. Barrett. New York, NY: Guilford Press.

Wang, L., G. McCarthy, A. W. Song, and K. S. LaBar. 2005. Amygdala activation to sad pictures during high-field (4 tesla) functional magnetic resonance imaging. *Emotion* 1:12–22.

Watanabe, S., and S. Masuda. 2010. Integration of auditory and visual information in human face discrimination in pigeons. Behavioral and anatomical study. *Behavioural Brain Research* 207:61–69.

Watanabe, S., U. Mayer, and H. J. Bischof. 2011. Visual Wulst analyses "where" and entopallium analyzes "what" in the zebra finch visual system. *Behavioural Brain Research* 222:51–56.

Watts, B. D., G. D. Therres, and M. A. Byrd. 2008. Recovery of the Chesapeake Bay bald eagle nesting population. *Journal of Wildlife Management* 72:152–58.

Weir, A. A. S., J. Chappell, and A. Kacelnik. 2002. Shaping of hooks in New Caledonian crows. *Science* 297:981.

Weir, A. A. S., B. Kenward, J. Chappell, and A. Kacelnik. 2004. Lateralization of tool use in New Caledonian crows (*Corvus moneduloides*). *Proceedings of the Royal Society of London B* 271:S334–36.

Weisskopf, M. G., E. P. Bauer, and J. E. LeDoux. 1999. L-Type voltage-gated calcium channels mediate NMDA-independent associative long-term potentiation at thalamic input synapses to the amygdala. *Journal of Neuroscience* 19:10512–19.

Welty, J. C., and L. Baptista. 1988. *The Life of Birds,* 4th ed. New York, NY: W. B. Saunders.

White, S. A. 2001. Learning to communicate. *Current Opinion in Neurobiology* 11:510–20.

Whiten, A. 2000. Primate culture and social learning. *Cognitive Science* 24:477–508.

Whittow, G. C., editor. 2000. *Sturkie's Avian Physiology,* 5th ed.. San Diego, CA: Academic Press.

Wichelman, D. F. 2006. *Sir James Crow: A True Story*. Self-published.

Wiltschko, R., and W. Wiltschko. 2009. Avian navigation. *Auk* 126:717–43.

Winter, D. 2005. *Year of the Crow: Adventures on Salt Spring Island*. Self-published.

Woolfson, E. 2008. *Corvus, A Life with Birds*. London: Granta Books.

Wootton, S. 2006. Nature's journal. The *Olympian,* September 9.

Wright, P., and A. M. Spence. 1976. Changes in emotionality following section of the tractus occipito-mesencephalicus in the Barbary dove (*Streptopelia risoria*). *Behavioural Processes* 1:29–40.

Wright, T. F., E. E. Schirtzinger, T. Matsumoto, J. R. Eberhard, G. R. Graves, J. J. Sanchez, S. Capelli, H. Müller, J. Scharpegge, G. K. Chambers, and R. C. Fleischer. 2008. A multilocus molecular phylogeny of the parrots *(Psittaciformes)*: support for a Gondwanan origin during the Cretaceous. *Molecular Biology and Evolution* 25:2141–56.

Wyles, J. S., J. G. Kunkel, and A. C. Wilson. 1983. Birds, behavior, and anatomical evolution. *Proceedings of the National Academy of Science USA* 80:4394–97.

Yellowstone Science. 2004. Raven predation of eared grebes. *Yellowstone Science* 12(3).

Yin, R. K. 1969. Looking at upside-down faces. *Journal of Experimental Psychology* 81:141–45.

Yorzinski, J. L., and S. L. Vehrencamp. 2009. The effect of predator type and danger level on the mob calls of the American crow. *Condor* 111:159–68.

Yorzinski, J. L., S. L. Vehrencamp, A. B. Clark, and K. J. McGowan. 2006. The inflected alarm caw of the American crow: Differences in acoustic structure among individuals and sexes. *Condor* 108:518–29.

Young, L. J., and Z. Wang. 2004. The neurobiology of pair bonding. *Nature Neuroscience* 7:1048–54.

Zapka, M., D. Heyers, C. M. Hein, S. Engels, N-L Schneider, J. Hans, S. Weiler, D. Dreyer, D. Kishkinev, J. M. Wild, and H. Mouritsen. 2009. Visual but not trigeminal mediation of magnetic compass information in a migratory bird. *Nature* 461:1274–78.

Zinkivskay, A., F. Nazir, and T. V. Smulders. 2009. What-where-when memory in magpies (*Pica pica*). *Animal Cognition* 12:119–25.

Index

Page numbers in *italics* refer to illustrations.

About the Authors

John Marzluff is Professor of Wildlife Science in the School of Environmental and Forest Sciences at the University of Washington. His research has been the focus of articles in *The New York Times, National Geographic, Audubon, Boys' Life, The Seattle Times,* and *National Wildlife*. PBS's *Nature* featured his raven research in its 2001 production "Ravens" and featured his crow research in 2010 with the documentary film "A Murder of Crows." John has been a guest on NPR's *Diane Rehm* show, *The Jay Thomas Show*, and *Morning Edition* (www.npr.org/crows).

John has won writing and speaking awards, including the A. Brazier Howell, Board of Directors, and H. R. Painton awards from the Cooper Ornithological Society. He is a fellow of the American Ornithologists' Union. He speaks regularly at professional conferences across the country, including to the American Ornithologists Union, the Ecological Society of America, and the Cooper Ornithological Society, as well as keynotes birding festivals, informs local birding and civic clubs, and addresses special events such as the Ketchikan Museum Lecture, the Egan Lecture, the Beaty Museum Biodiversity Lecture, and the Athens Speaker Series.

His graduate (Northern Arizona University, studying with Professor Russell Balda) and initial postdoctoral (University of Vermont with Professor Bernd Heinrich) research focused on the social behavior and ecology of jays and ravens. He has authored more than one hundred scientific papers on various aspects of bird behavior and wildlife management in addition to the four books he's written.

Tony Angell has authored and/or illustrated a dozen award-winning books related to natural history. Most recently, his drawings in the coauthored book *In the Company of Crows and Ravens* received the prestigious Victoria and Albert Museum illustration prize. Angell's artwork also includes both stone and bronze sculptures, which are the basis for his one-man shows and retrospectives that he has regularly had over the past forty years. His works are continuously available at galleries in Seattle and Santa Fe and are included in several museum and corporate collections throughout the country, including the Gilcrease Museam of Art in Tulsa, Oklahoma, the Frye Art Museum of Seattle, and the National Museum of Wildlife Art in Jackson, Wyoming. He has served as chairman and regular board member of Washington State's chapter of the Nature Conservancy, and for his work in conservation he received the Nature Conservancy's highest award, the Gold Oakleaf from the national office as well as the Finestone Award from New York University at Stonybrook. For more than thirty years, Angell supervised the State Office of Environmental Education in Washington and developed a wide range of information and curriculum related to a broad spectrum of environmental issues. Angell is an elected member of the American Ornithologists Union and an elected fellow of the National Sculpture Society and the Explorers Club.